CIRCADIAN RHYTHMS OF CARDIOVASCULAR DISORDERS

edited by

Prakash C. Deedwania, MD

Chief, Cardiology Division
V.A. Medical Center
University of California San Francisco
Program at Fresno, California;
Clinical Professor of Medicine
University of California San Francisco School of Medicine
San Francisco, California

FUTURA

Futura Publishing Company, Inc.
Armonk, NY

Library of Congress Cataloging-in-Publication Data

Circadian rhythms : cardiovascular implications / edited by Prakash C.
   Deedwania.
         p.   cm.
      Includes bibliographical references and index.
      ISBN 0-87993-632-0
      1. Heart—Pathophysiology.   2. Circadian rhythms.   I. Deedwania,
   Prakash C., 1948-   .
      [DNLM: 1. Cardiovascular Diseases—physiopathology.   2. Circadian
   Rhythm—physiology.   WG 120 C578 1997]
   RC682.9.C55   1997
   616.1'07—dc21
   DNLM/DLC
   for Library of Congress   96-44295
                                                                        CIP

Copyright 1997
Futura Publishing Company, Inc.

Published by
Futura Publishing Company, Inc.
135 Bedford Road
Armonk, NY 10504

LC#: 96-44295
ISBN#: 0-87993-632-0

Printed in the United States of America
This book is printed on acid-free paper.

To my lovely wife, Cathy,
and my children,
Anne, Ravi, and David,
for their patience, understanding,
continued support and inspiration

Contributors

Antonio Bayés de Luna, MD
Professor of Medicine, Cardiology, Autonomous University of Barcelona; Director, Department of Cardiology and Cardiac Surgery, Hospital Sant Pau, Barcelona, Spain

Simon Chakko, MD
Associate Professor of Medicine, University of Miami School of Medicine; Chief, Non-invasive Cardiac Laboratories, V.A. Medical Center, Miami, Florida

Juan Cinca, MD
Cardiology Service, Hospital Vall d'Hebró, Barcelona, Spain

Prakash C. Deedwania, MD
Chief, Cardiology Division, V.A. Medical Center, University of California San Francisco Program at Fresno, California; Clinical Professor of Medicine, University of California San Francisco School of Medicine, San Francisco, California

Charles R. Lambert, MD, PhD
Abraham Mitchell Professor of Medicine; Director, Cardiac Catheterization Laboratory, University of South Alabama, Mobile, Alabama

John Marler, MD
Medical Officer, Public Health Service, National Institutes of Health, National Institute of Neurological Disorders and Stroke, Bethesda, Maryland

Angel Moya, MD
Cardiology Service, Hospital Vall d'Hebró, Barcelona, Spain

v

David Mulcahy, MD, MRCPI, FESC
Cardiology Branch, National Heart, Lung, & Blood Institute, National Institutes of Health, Bethesda, Maryland

James E. Muller, MD
Professor of Internal Medicine; Chief, Division of Cardiology, University of Kentucky, Lexington, Kentucky

Jaime E. Murillo, MD
Post Doctoral Associate, Institute for Prevention of Cardiovascular Disease, Cardiovascular Section, Yale-New Haven Hospital, New Haven, Connecticut

Robert J. Myerburg, MD
Professor of Medicine and Physiology; Director, Division of Cardiology, University of Miami School of Medicine, Miami, Florida

Robert W. Peters, MD
Professor of Medicine, The University of Maryland School of Medicine; Chief, Division of Cardiology, The Baltimore Veterans Administration Medical Center, Baltimore, Maryland

Arshed A. Quyyumi, MD, FACC
Senior Investigator, Cardiology Branch, National Heart, Lung & Blood Institute, National Institutes of Health, Bethesda, Maryland

Michael H. Smolensky, PhD
Director, Hermann Center for Chronobiology and Chronotherapeutics; Professor, University of Texas, Schools of Public Health & Medicine, Health Sciences Center, Houston, Texas

Geoffrey H. Tofler, MD
Associate Professor of Medicine, Institute for Prevention of Cardiovascular Disease, Cardiovascular Division, Deaconess Hospital, Harvard Medical School, Boston, Massachusetts

Xavier Viñolas, MD
Cardiology Department, Hospital Sant Pau, Barcelona, Spain

Robert G. Zoble, MD, PhD
Assistant Professor of Medicine/Cardiology, University of South Florida College of Medicine, Medical Director, Coronary Care Unit, James A. Haley Veterans' Hospital, Tampa, Florida

Foreword

Nothing is more certain than the rising and setting of the sun during each 24-hour period. This daily pattern of light and dark and our own daily patterns of being awake and asleep are as fundamental to our understanding of our own biology as any other process. And yet it has been only in the past decade that we have really understood how important this cycle is in relationship to disease, particularly cardiovascular disease. The morning predominance of myocardial infarction, myocardial ischemia, and sudden death have now been well documented in a number of large population studies. This fact provides the stimulus for better understanding our own chronobiology and the periodicity of circadian rhythms which contribute to this predominance of adverse events in the morning hours. People who thought that Monday morning was a particularly difficult time also appear to have more justification for this belief than they realized. There is now a burgeoning literature on potential mechanisms for these circadian patterns of adverse events in coronary disease. A number of mechanisms may play important roles—from changes in heart rate, blood pressure, catecholamines, and other neurohormones to changes in clotting, electrophysiology, and vascular reactivity.

This book *Circadian Rhythms of Cardiovascular Disorders,* edited by Dr. Prakash Deedwania, summarizes all of this important information in one volume. This is a must read for clinicians who care for patients with heart disease. An understanding of these patterns and their pathophysiology are important for understanding both the timing and the type of therapy that can prevent such adverse events. This book has been written by experts in the field. Each one has had a special interest in the subject of his chapter and is able to summarize the relevant information in a concise manner. Altogether, this is an exciting and informative book, which is the first of its kind. This book will stand as a milestone on information related to circadian patterns. Although we may refine our knowledge of these patterns in the future in ways that are not yet evident, it is clear that this book will set the stage for all of the research that is to follow. Dr. Deedwania and the authors of all of the chapters are to be congratulated for providing this important information to all physiologists and clinicians interested in cardiovascular disease.

William W. Parmley, MD
Chief of Cardiology
H. C. Moffitt/Joseph Long Hospitals
Professor of Medicine
University of California, San Francisco, Medical Center
San Francisco, California

Preface

Grim death took me without warning
I was well at night
and dead at nine in the morning
Epitaph, Sevenoaks, England

For centuries, anecdotal experiences such as this one described in the epitaph found in the Sevenoaks Churchyard, England, have been cited to illustrate the increased risk of death in the morning hours. Despite this knowledge, it was not until the past decade that the circadian pattern of acute myocardial infarction and sudden cardiac death was established by observations from large, epidemiological studies. Subsequently, a large number of well-designed studies published in the last 8–10 years have demonstrated that, in addition to acute myocardial infarction and sudden cardiac death, episodes of myocardial ischemia, cardiac arrhythmias, and cerebrovascular accidents also have a well-defined circadian pattern. Most of these studies have defined a surge in the risk of cardiovascular events during the morning hours. Recent studies have demonstrated that not only is there a daily cyclical pattern but also a weekly, a monthly, and a seasonal variability in cardiovascular events.

Although in the beginning the literature about circadian rhythms of cardiovascular disorders provided intriguing data, it was not until recently that the pathophysiological processes responsible for the circadian periodicity were well defined. We now understand that a number of important physiological parameters such as heart rate, blood pressure, vascular reactivity, cardiac contractility, and various hemostatic factors all demonstrate a circadian pattern similar to that described for cardiovascular disorders. Several recent studies have also emphasized the importance of changes in posture, time of awakening, physical activity, mental stress, and anger, etc., as potential triggers that might act in concert with the chronological changes in the hemodynamic and hemostatic factors. Although there has been some delay, the chronological pattern of cardiovascular disorders and the etiological role of potential triggers are now widely accepted. The clinicians have now begun to realize the importance of these data in relationship to the prevention of the onset of cardiovascular events. In addition, chronopharmacological therapeutic systems have recently become available with the potential of delivering drugs at times of the greatest risk of cardiovascular events. It is, therefore, now imperative for clinicians to be aware of all important data in this burgeoning area of cardiovascular disorders. It is with this need in mind that I have brought together the leading experts in the field to provide a comprehensive and authoritative review of this subject.

Circadian Rhythms of Cardiovascular Disorders is the first book of its kind in the medical literature and provides the clinician with a compendium of articles by pioneers in the field. This book offers the reader the most up-to-date literature with a careful review of the epidemiological and scientific evidence supporting the circadian rhythmicity of cardiovascular disorders. The contributing authors have taken great care in describing the clinical relevance of the data, including detailed discussion about the pathophysiological processes and potential therapeutic considerations for the practicing physician. The first four chapters of the book describe the role of triggers and other modulating factors responsible for circadian periodicity of cardiovascular events. These chapters are followed by a series of chapters that describe circadian patterns of myocardial ischemia, acute myocardial infarction, cardiac arrhythmias, sudden cardiac death, and cerebrovascular accidents. The last section of the book includes chapters specifically dealing with the recent advances in the field of chronopharmacology and therapeutic approaches designed to prevent the onset of cardiovascular disorders. All chapters have been authored by leading experts in the field who have provided concise, practical, and authoritative reviews of the subject. A serious, concerted effort has been made to provide the reader with all of the relevant material in a well-coordinated fashion, making this book more than just a compendium of articles in the field. It is my sincere hope that this book will provide the clinician with meaningful, up-to-date information in this rapidly expanding and exciting field of cardiology.

I would like to express my sincere gratitude to the world-class leaders in the field for their invaluable contributions. I would also like to acknowledge all the help given by Ms. Laurie Harrington for her valuable assistance assembling and keeping this book in a timely fashion. Finally, I would like to thank my wife and children who patiently endured through all the time I took away from them and who were a constant source of inspiration in my efforts in putting this book together.

Prakash C. Deedwania, MD

Contents

1

Circadian Patterns of Acute Myocardial Infarction:

The Role of Triggers and Other Modulating Factors

Robert G. Zoble, MD, PhD,
Prakash C. Deedwania, MD

Introduction

Circadian patterns have been observed for a variety of cardiovascular disorders, including cardiac arrhythmias,[1–6] sudden cardiac death,[7,8] cerebrovascular events,[9,10] episodes of stable angina,[11,12] unstable angina,[13] and acute myocardial infarction (MI).[14–40]

The circadian timing of onset of acute MI provides important insights into factors triggering or modifying the myocardial infarction process. Most studies have detected circadian periodicity that is nonuniform and statistically significant [14,15,18,19,21–23,26,27,29,31–34,36,37,39,40] (Fig. 1). However, a few investigators have either failed to observe significant circadian variation[16,25,30] or did not perform appropriate statistical analyses to convincingly demonstrate the circadian pattern in their study population[20,24,28,35] (Fig. 2). However, in some of the studies (e.g.,

From: Deedwania PC (ed): *Circadian Rhythms of Cardiovascular Disorders.*
©Futura Publishing Co., Inc., Armonk, NY, 1997.

Circadian Pattern of Acute MI: Significant Periodicity

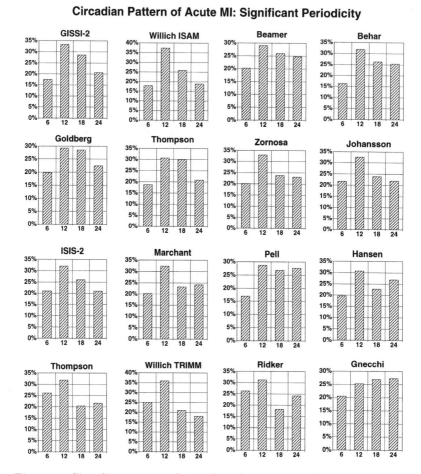

Figure 1: Circadian patterns for studies observing significant periodicity.

WHO) where statistical analyses were not performed, a nonuniform circadian pattern appears to be present.[35]

In the studies demonstrating significant circadian periodicity, differences have been noted in the timing of the peak incidence of MI onset. Most studies have observed peak incidence to occur in the second quarter of the day from 6:01 AM to 12:00 noon,[14,15,18,21–23,27,29,32–34,36,37,40] but a fourth quarter peak (6:01 PM to midnight) has also been reported.[19] In some instances, two peaks have been noted. Thompson observed equal second and third quarter peaks,[33] Gnecchi observed equal third and fourth quarter peaks,[19] and Hansen reported a major second quarter peak with a minor fourth quarter peak.[21]

Circadian Pattern of Acute MI: No Significant Periodicity

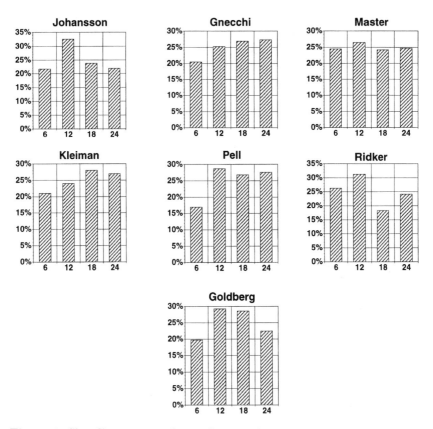

Figure 2: Circadian patterns for studies not observing significant periodicity.

In contrast to the reported differences in the peak incidence of myocardial infarction onset, there is greater agreement on the timing of the nadir (low point) of MI onset. The nadir has been observed in the first quarter of the day (12:01 AM to 6:00 AM) in the majority of reports.[14–16,18,19,21–23,25,27–29,33,36,40] A few authors have reported the nadir to occur at other times—either in the third quarter[32,34] or in the fourth quarter[37]—but never in the second quarter. A double nadir has been noted in the ISIS-2 data (first quarter and fourth quarter equally low),[23] and has also been reported by Hjalmarson (first quarter and third quarter equally low),[22] by Muller (first quarter and fourth quarter equally low),[29] and by Willich in the ISAM study (first quarter and fourth quarter equally low).[36]

Overall, the most typical circadian pattern of myocardial infarction has a nadir in the first quarter, a peak in the second quarter, followed by intermediate values in the third and fourth quarters. This pattern has been demonstrated in the vast majority of studies reporting MI timing for a large cohort of patients. It is, therefore, reasonable to assume that the typical circadian pattern of MI represents a first quarter nadir followed by a second quarter peak when evaluating the role of various triggers and modifying factors.

Triggers and Circadian Pattern

The pathophysiological mechanisms underlying the circadian periodicity of acute MI onset are complex and relate to interactions between protective elements and triggering factors, some of which may be intrinsic and some extrinsic. In fact, some of the reported differences in observed circadian patterns may indeed relate to varying degrees of influence of the protective and triggering factors during different time periods in various patient populations.

In most patients, the first quarter of the day is usually associated with sleep, when heart rate and blood pressure are at their daily minimums.[41–43] Additionally, there are no external stresses during sleep. Thus, at least from the standpoint of physical and psychological stress, it would be reasonable to expect that the contribution of factors responsible for MI onset would be minimal in the first quarter of the day and, indeed, most studies have reported this to be the case.

However, with awakening from sleep and resumption of upright posture in the second quarter of the day, several possible triggers for MI are brought into play. Rising to an upright posture leads to gravitational blood pooling in the lower extremities and reductions in left ventricular dimension and stroke volume. The associated fall in blood pressure activates the sympathetic nervous system, which in turn triggers compensatory increases in heart rate and blood pressure. If patients with fixed coronary obstructions cannot meet the increased oxygen demands associated with increased heart rate and blood pressure, myocardial ischemia or an acute MI may result. The outstripping of oxygen supply, if sustained, may be an important mechanism responsible for non-Q-wave MI, since this type of MI (non-Q MI) is generally not associated with total thrombotic coronary occlusion.[44]

In contrast to non-Q-wave MI, a Q-wave MI is typically associated with total thrombotic coronary occlusion. Coronary angiography in the first 4 hours of Q-wave MI has demonstrated that 85% of patients have evidence of total thrombotic coronary occlusion.[45] Therefore, triggering mechanisms for non-Q-wave MI and Q-wave MI may be quite different. Because coronary thrombosis is responsible for many acute MIs, further consideration and evaluation of the circadian patterns of factors promoting or inhibiting thrombosis may provide important insights into the triggering mechanisms of MI.

The Role of Sympathetic Activation as a Potential Trigger

As a consequence of the increased physical demands placed on the heart by arousal, the sympathetic nervous system is activated. Increased blood pressure and its first derivative (dp/dt) with increased shear force may lead to intracoronary plaque rupture or fissuring. The intracoronary raw surface of the fissured/ruptured plaque then becomes a nidus for development of an occlusive coronary thrombus. This mechanism becomes even more understandable when one realizes that there are other thrombogenic factors also operating during the second quarter of the day.

The second quarter thrombogenic milieu is created by changes in various hormonal endocrine and hematologic factors that favor thrombosis or reduce intrinsic fibrinolytic activity. Catecholamines released from the adrenal glands and from sympathetic neurons may indeed play an important role in this regard. Circadian patterns of plasma epinephrine and norepinephrine demonstrate nadirs from 11:00 PM to 5:00 AM for epinephrine and from 1:00 AM to 9:00 AM for norepinephrine.[46] While catecholamines are lowest in the first quarter of the day, their levels peak in the late morning and early afternoon. The circadian pattern of catecholamine activity is further supported by the fact that urinary excretion of epinephrine and norepinephrine and their metabolites is also lowest from midnight to 6:00 AM and usually peaks from 6:00 AM to noon.[17]

Thus, plasma and urinary catecholamine levels demonstrate circadian periodicity consisting of a first quarter trough followed by a second quarter increase. As previously noted, catecholamines can promote triggering of coronary events by placing increased physical demands on the heart, both directly (positive inotropic state) and indirectly (increased peripheral vascular resistance). Catecholamines can also contribute to triggering of MI by a second mechanism, by increasing coronary vasomotor tone, and in some cases leading to frank coronary vasospasm. This could lead to decreases in coronary flow or in some cases actual flow cessation and stasis for the duration of vasospasm, which could promote intracoronary thrombosis.

Sympathetic activation and increased catecholamine activity can promote MI in the second quarter by yet a third mechanism: enhancing platelet aggregation. Tofler and coworkers measured the circadian pattern of platelet aggregation in response to adenosine diphosphate (ADP) and epinephrine.[47] Platelet aggregation was lowest from 3:00 AM to 6:00 AM (before arising), but then peaked at 9:00 AM (60 minutes after arising). This difference was statistically significant. In a second experiment, Tofler and associates demonstrated that subjects remaining in bed until after 9:00 AM did not show the 9:00 AM increase in platelet aggregation. Thus, the second quarter increase in platelet aggregation was shown to be related to postural change rather than to the time of day per se. Several previous studies have shown that MI onset is also closely linked to the time of awakening/arising, the risk of developing an infarction being greatest in the first 3 hours after awakening.[37,48]

Thus, it appears that sympathetic activation in the second quarter of the day can trigger MI by four mechanisms: (1) increased cardiac demands, which may outstrip oxygen supply, (2) increased cardiac inotropy, which may cause plaque rupture, (3) increased coronary resistance, which may promote decreased oxygen delivery and thrombosis due to decreased flow, and (4) increased platelet aggregation.

The Thrombogenic Milieu and Circadian Pattern

In addition to increased platelet aggregability during the second quarter of the day (6:00 AM to noon), there is a thrombogenic milieu during this period due to a lowered intrinsic thrombolytic state[49] and higher levels of fibrinogen.[50] Andreotti measured the circadian activity of tissue-type plasminogen activator (t-PA) and its inhibitor, plasminogen activator inhibitor (PAI). He observed markedly reduced activity of the intrinsic t-PA during the early morning hours from 3:00 AM to 6:00 AM, which was associated with a peak in the activity of its inhibitor (PAI). The activity of t-PA increased slightly at 9:00 AM but did not peak until 6:00 PM, so that from 6:00 AM to noon intrinsic fibrinolytic activity was well below peak values. This lowered intrinsic fibrinolytic activity preceding and during the second quarter of the day might indeed contribute to the increased risk of MI in the second quarter.

From this discussion, it is clear that sympathetic activation associated with the process of awakening and rising to an upright posture in the second quarter of the day plays a major role in triggering MI. The mechanisms involved include increased physical demands on the heart at a time when coronary flow may be decreased and a thrombogenic milieu also exists. The thrombogenic milieu during this period is due to increased levels of catecholamines, increased platelet aggregability, and decreased intrinsic fibrinolytic activity.

Other factors may also be responsible for the circadian pattern of MI and these might act as triggers or protective factors and are worthy of discussion. Several studies have examined cardiovascular medications, age, gender, or medical conditions promoting coronary artery disease or events related to it.[14–40] Medications studied have included aspirin, beta blockers, and calcium channel blockers. Various disease processes or coronary artery disease risk factors that have been evaluated include: smoking, age, gender, prior MI, diabetes, hypertension, prior congestive heart failure, prior cerebrovascular accident, peripheral vascular disease, and prior angina. Additionally, non-Q-wave MI has been compared to Q-wave MI. These subgroups are worth considering in detail to gain further insights into the pathophysiological processes responsible for MI.

The Role of Aspirin in Circadian Patterns

Aspirin inhibits platelet aggregation, and in two studies it has been shown to be effective in primary[51] and secondary[52] prevention of acute MI. The influence of aspirin on the circadian pattern of acute MI has been investigated by Ridker[32] and by the ISIS-2 investigators[23] (Fig. 3). Ridker reported that aspirin abolished the circadian periodicity of MI in a small study (n=211), while the much larger ISIS-2 study (n=12,163) failed to observe an effect of aspirin on the circadian pattern of MI. However, patients in the study by Ridker et al. assigned to

Circadian Pattern of Acute MI: ASPIRIN

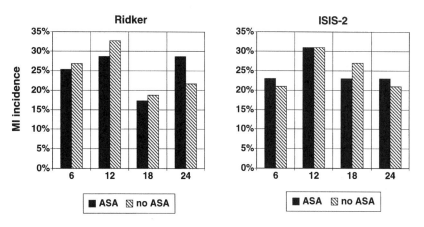

Figure 3: The effect of aspirin (ASA) on circadian patterns of acute MI.

aspirin were all taking 324 mg of aspirin every other day, whereas in the ISIS-2 study, patients taking any dose of aspirin in the prior week were assigned to the aspirin subgroup. The observations by Ridker et al. are consistent with aspirin's antiplatelet effects, but suffer due to small numbers. The failure of the ISIS-2 study to observe an aspirin-mediated blunting of the circadian pattern of MI may well have been due to suboptimal aspirin dosing. It will be important for future studies evaluating the effects of aspirin on the circadian pattern of coronary events to have larger numbers of patients with precise data on aspirin dosing.

The Effects of Beta Blockers on Circadian Patterns

The circadian pattern of acute MI in patients previously on beta blockers has been examined in five studies[21,22,25,29,36] (Fig. 4). In studies by Muller et al.[29] and Willich and associates,[36] the overall MI group demonstrated significant circadian periodicity that was not present in patients who were on beta blockers prior to MI. Hansen and coworkers[21] observed a more complex phenomenon and showed that prior use of nonselective beta blockers abolished circadian periodicity of MI, but the use of cardioselective beta blockers failed to do so. Hjalmarsen et al.[22] also observed that prior beta blocker use altered the circadian pattern of MI and showed that those not taking beta blockers experienced a second quarter single peak, while those with prior use of beta blockers had dual peaks in the second and fourth quarters of the day. Although the differences between the subgroups taking or not taking beta blockers were not statistically significant (p=0.10), those receiving beta blockers had a smaller second quarter peak (26%) compared to those not taking beta blockers (28.4%). Additionally, Kleiman examined the circadian effects of beta blockers in a pure non-Q-wave MI population. No circadian periodicity was observed in the overall group and no differences were seen in the subgroups taking or not taking beta blockers. Despite failing to observe statistically significant circadian effects of beta blockers in some of these studies, several studies have demonstrated reductions in the incidence of MI in the second quarter of the day.[21,22,25,29,36]

The alterations observed in the circadian pattern of MI produced by beta blockers are consistent with their ability to blunt sympathetic activation. Beta blockers might also antagonize catecholamine-mediated platelet aggregation, but the clinical significance of this mechanism has been called into question, at least for patients with cardiac is-

Circadian Pattern of Acute MI: BETA BLOCKERS

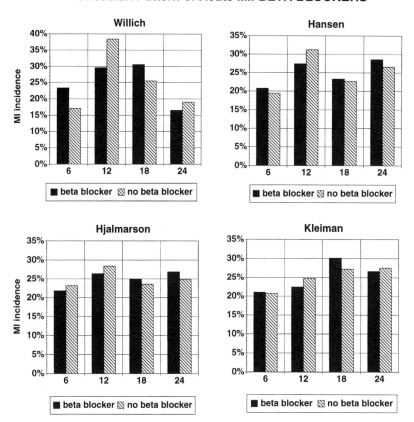

Figure 4: The effect of beta blockers on circadian patterns of acute MI.

chemia.[53] Willich studied the effects of beta blocker therapy (metoprolol 200 mg bid) on the circadian pattern of platelet aggregation and silent myocardial ischemia in 10 patients with known coronary artery disease. He observed metoprolol-mediated blunting of the circadian peak of silent ischemia on ambulatory monitoring, but without blunting of the circadian peak of platelet aggregation. This observation might suggest that the role of sympathetic activation in triggering myocardial ischemia is related primarily to increased oxygen demand in the face of limited coronary reserve rather than being secondary to increased platelet aggregation. It is important to note, however, that the mechanisms of transient myocardial ischemia and non-Q-wave MI may be different from those of transmural MI.

The Effects of Calcium Channel Blockers

The effects of calcium channel blockers on the circadian pattern of MI have been examined by Hansen et al.[21] and by Willich and associates[36] (Fig. 5). Hansen and coworkers observed that patients already on calcium channel blockers prior to MI onset had a second quarter circadian peak that was significantly nonuniform (p<0.01). The circadian pattern of MI in the calcium channel blocker subgroup was similar to that of the overall group. Willich et al. also observed the circadian pattern of MI in those on calcium channel blockers with a significant second quarter peak similar to that of the overall group. Thus, in contrast to beta blockers, calcium channel blockers have not been associated with modification of the circadian periodicity of MI. However, it is important to note that these were analyses of pooled calcium channel blocker populations. No comparisons have been made between dihydropyridine calcium channel blockers and nondihydropyridine calcium channel blockers (e.g., verapamil and diltiazem). Since beta blockers directly blunt sympathetic activation while calcium channel blockers do not, the differing effects of these two classes of drugs on circadian pat-

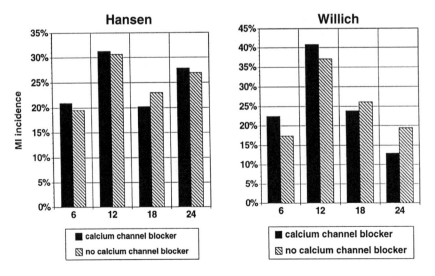

Circadian Pattern of Acute MI:
CALCIUM CHANNEL BLOCKERS

Figure 5: The effect of calcium channel blockers on circadian patterns of acute MI.

tern of MI further support the role played by sympathetic activation in MI triggering and periodicity. Clearly, more work is needed in this area to prospectively examine the role of different classes of calcium channel blockers as well as the role of unique timed-release preparations.

The Role of Smoking

The influence of smoking on the circadian pattern of acute MI has been examined in three studies[18,21,22] (Fig. 6). The GISSI-2 investigators[18] observed smokers to have a significantly blunted circadian pattern (p<0.001) compared to nonsmokers. However, Hjalmarsen et al.[22] observed a significantly nonuniform (p<0.001) circadian pattern in the overall group, with two peaks—a large one in the second quarter and a lesser one in the fourth quarter. In comparison, the nonsmoker subgroup demonstrated a single second quarter peak. Hjalmarsen et al. compared smoking and nonsmoking subgroups and found smoking to be a significant modifier (p<0.03) of the circadian pattern of MI. Hansen and coworkers[21] observed a significantly nonuniform (p<0.001) circadian pattern in the overall group, but with two peaks—a large one in the second quarter and a lesser peak in the fourth quarter. The circadian pattern was preserved in smokers and was also significantly nonuniform (p<0.001). However, Hansen et al. did not statistically compare the smoking and nonsmoking subgroups to one another.

Thus, all three studies observed smoking to blunt the circadian periodicity of MI. The reason for this is not clear. Smoking increases blood pressure, causes vasoconstriction, and increases platelet aggregation,

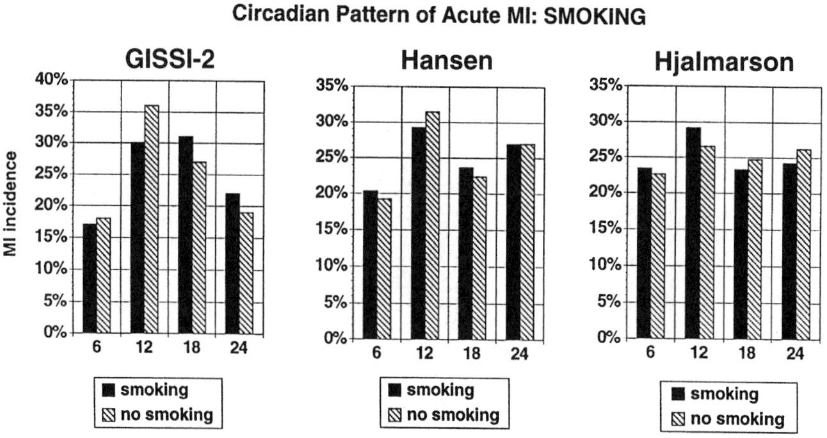

Figure 6: The effect of smoking on circadian patterns of acute MI.

and thus smoking might be expected to increase the risk of MI during its peak in the second quarter because most people smoke cigarettes in the morning after awakening. However, exactly the opposite occurs. It may be that smoking increases the incidence of MI throughout the day, thereby blunting the circadian pattern and the increased risk of MI in the morning hours.

Effect of Age on Circadian Pattern of MI

Six studies have examined the role age plays in the circadian pattern of MI[15,16,18,19,21–23] (Fig. 7). Subgroup comparisons have used varying age cutpoints of 70 years (three studies) and 65 years (two studies), one study examined four age groups. In all six studies, peak MI inci-

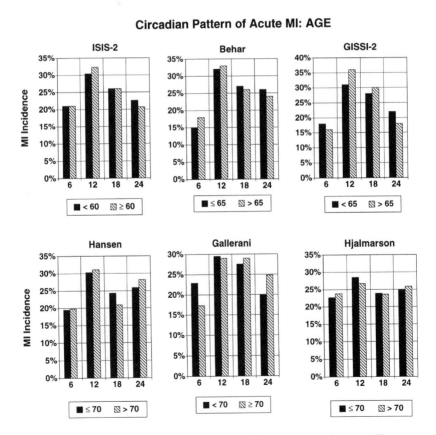

Figure 7: The effect of age on circadian patterns of acute MI.

dence for the overall groups occurred in the second quarter, with the lowest incidence in the first quarter.

Hansen et al.[21] observed a significantly nonuniform (p<0.001) circadian pattern in the overall group with a second quarter peak. This pattern was also significant in those aged >70 years (p<0.001) and in those ≤70 years (p<0.001). However, in comparing the two age subgroups directly with one another, those >70 years differed significantly from those ≤70 years by demonstrating a second peak in the fourth quarter (p=0.007, test of independence). The magnitude of the second quarter peak, however, was similar in both age subgroups. Hjalmarsen and associates[22] observed a significantly nonuniform (p<0.001) circadian pattern with a second quarter peak. Those aged ≤70 years had a single prominent second quarter peak, while those >70 years had a smaller second quarter peak, but the older patients had a fourth quarter peak as well. However, the differences between the two age subgroups were not statistically significant. Thus, both Hansen and Hjalmarsen observed advanced age to be associated with an increased incidence of MI in the fourth quarter. The reason for this observation is not clear, although a secondary peak in MI incidence in the evening has been reported by others also and may possibly relate to the evening meal.[48]

The GISSI-2 investigators[18] observed a significant (p<0.001) second quarter peak in the overall group. Compared to those <65 years of age, the magnitude of the second quarter peak was significantly greater (p<0.001) in those ≥65 years of age (36.3% vs. 31.1%). Thus, the GISSI-2 investigators found advanced age to be associated with an enhanced second quarter peak. This differs from the increased fourth quarter MI incidence seen by Hansen et al. and Hjalmarsen and associates in those with advanced age. While these three studies observed that advanced age increased MI incidence in the second or fourth quarter, other studies have failed to observe the influence of age on the circadian pattern of MI.[15,16,23]

The ISIS-2 investigators[23] observed a second quarter peak that was significantly greater (p<0.001) than uniformity would anticipate in the overall group and in all four age groups examined. All four age subgroups had circadian patterns similar to the overall group pattern. However, no direct comparisons between various age groups were made.[23] Behar et al. observed a circadian pattern with a second quarter peak that was significantly nonuniform (p<0.01) in the overall group.[15] Circadian patterns of those ≤65 years of age and those >65 years were similar to each other and the overall group.[15] The subgroup statistical comparisons, however, were not reported. Gallerani et al. also found no significant circadian pattern in the overall group or in four different age groups.[16]

The reasons for differing circadian patterns in various age subgroups are not clear and may be multifactorial, involving different prevalence of modifying factors (such as smoking or medications) in people of different ages. The fact that this may be the case is illustrated by the GISSI-2 results. The GISSI-2 investigators found advanced age to be associated with an enhanced second quarter peak. However, this enhancing effect might have been related not directly to older age per se, but to another modifying variable (e.g., smoking). In the GISSI-2 study, nonsmokers had a higher second quarter peak than smokers. In this study 36% of the younger age group were nonsmokers, while 71% of patients older than 65 years were nonsmokers. If smoking blunts the circadian pattern, then a subgroup with a lower percentage of smokers (older patients) will have a relatively less blunted circadian pattern. Therefore, in comparison,the subgroup with fewer smokers will have a comparatively enhanced circadian pattern, even if there is no intrinsic difference in the subgroups due to age.

Gender and Circadian Rhythmicity

Five studies have examined the role gender plays in circadian patterns of MI[15,16,18,21-23] (Fig. 8). As would be expected from studies involving acute MI, the majority of patients in the gender subgroup analyses were males (average 73%; range 66%–76%). The circadian patterns of MI in the two genders were similar in the studies of Behar et al.,[15] Hansen et al.,[21] and the ISIS-2 study.[23] Hjalmarsen et al.[22] observed a significantly nonuniform dual peak circadian pattern in the overall group and in females. Males demonstrated a single second quarter peak.[22] However, the gender subgroup patterns were not statistically different from each other. Gallerani et al. observed a second quarter peak in the overall group and in females, but a third quarter peak in males.[16] In the report by Gallerani et al., however, none of the peaks, either in the overall group or in either of the genders, were significant.[16] Thus, while significantly nonuniform circadian patterns have been demonstrated for both males and females, no significant difference in circadian patterns of MI has been demonstrated between the genders. It would therefore be reasonable to conclude that both males and females appear to be equally prone to the effects of various triggers of MI.

Non-Q-Wave MI

The majority of the studies examining the circadian pattern of acute MI have reported on data pooling Q-wave and non-Q-wave MIs. There have been only three reports that specifically address the issue

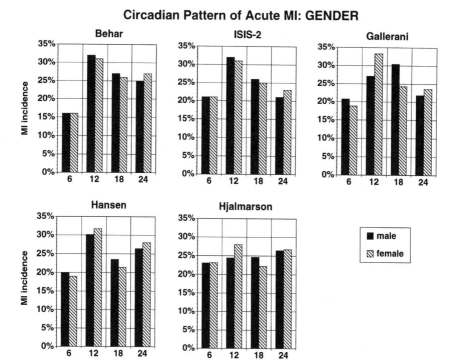

Figure 8: The effect of gender on circadian patterns of acute MI.

of the circadian pattern of non-Q-wave MI[21,22,25] (Fig. 9). Hjalmarson et al.[22] and Hansen et al.[21] performed non-Q-wave MI subgroup analyses on pooled MI data, while Kleiman et al.[25] reported the experience from the purely non-Q-wave MI patients enrolled in the Diltiazem Reinfarction Study (DRS). These three studies have reported that the circadian pattern of the non-Q-wave MI differs from that most typically seen for pooled MI data (which consist primarily of Q-wave MIs).

Hjalmarsen et al. observed peak onset of MI to be in the second quarter of the day from 6 AM to noon with 28% of the population developing symptoms during that time period.[22] This circadian pattern was significantly nonuniform (p<0.001). In contrast, patients with non-Q-wave MI (n=832) had their peak MI incidence (28.1%) in the evening (fourth quarter). Thus, according to the observations of Hjalmarsen et al., the circadian pattern of non-Q-wave MI differs qualitatively from that of Q-wave MI in that the risk of Q-wave MI peaks in the second quarter of the day whereas the incidence of non-Q-wave MIs peaks in the fourth quarter.[22]

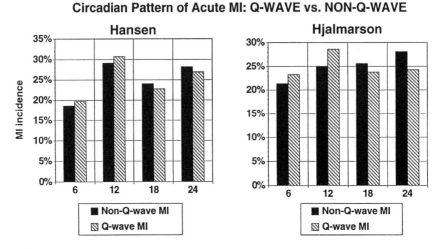

Figure 9: Non-Q-wave versus Q-wave MI and circadian patterns of acute MI.

Hansen and coworkers[21] observed a significantly ($p<0.001$) non-uniform circadian pattern in the overall group with dual peaks in the second quarter (30.7%) and in the fourth quarter (26.9%) of the day. A qualitatively similar bimodal circadian pattern was seen in non-Q-wave MI patients (n=333), which was also significantly ($p<0.05$) nonuniform. The second quarter peak was somewhat smaller (29.1%) in the non-Q-wave MI group than in the pooled MI group. Thus, from a qualitative standpoint, Hansen and associates observed the circadian pattern of non-Q-wave MI to parallel that of Q-wave MI.[21] This is in contrast to the findings described by Hjalmarsen. In part, the difference between the results of these two studies might well relate to differences in the inclusion criteria used to define non-Q-wave MI diagnosis in the two studies.[21,22] In the study by Hansen et al.,[21] less than 5% of the population was considered to have non-Q-wave MI, whereas in the study by Hjalmarsen et al. (22), this percentage was more than three times larger (17.3%).

In the report by Kleiman et al. of a pure non-Q-wave MI population, no circadian pattern was detected for the enrolled or screened patients when examining the patterns in either 2-hour or 6-hour intervals.[25] However, examination of 8-hour intervals did reveal a significant (p=0.02) nadir between midnight and 8:00 AM (27%), compared to the 37% incidence between 8:00 AM and 4:00 PM and the 36% incidence between 4:00 PM and midnight.[25] Thus, the 8-hour data analysis revealing a nadir for MI onset during sleeping hours appear to be similar to that seen in most pooled MI studies. The authors concluded that in contrast to Q-wave MI, there is no preponderance of non-Q-wave infarction in the late morning, suggesting that the pathogenesis of the non-Q-wave MI might be different from that of transmural MI.

Thus, the three trials examining the circadian pattern of non-Q-wave MI have produced disparate results.[21,22,25] Hjalmarsen et al. observed a significantly nonuniform pattern, with a fourth quarter peak.[22] Hansen and associates also observed a significantly nonuniform pattern, but with a primary second quarter peak and a secondary fourth quarter peak.[21] In contrast to these two studies, Kleiman et al. did not observe a significant non-Q-wave MI circadian pattern.[25] Not only did these three studies differ with regard to the circadian pattern of non-Q-wave MI, they also differed with regard to concordance of non-Q-wave and Q-wave MI circadian patterns within the same study population. Hansen et al. observed a similar circadian pattern for his non-Q-wave and Q-wave MI subgroups.[21] In contrast, the subgroups in the study by Hjalmarsen et al. differed, with Q-wave MI patients showing a second quarter peak, but non-Q-wave MI patients having a fourth quarter peak.[22]

Further studies will be needed to clarify the circadian pattern of non-Q-wave MI and to determine if it is similar or dissimilar to that of Q-wave MI. Comparing future studies to the reported literature will have to take into account the increasing use of thrombolytic therapy. Thrombolytics can abort Q-wave MIs and convert them into non-Q-wave MIs. This may indeed modify the observed circadian patterns of non-Q-wave MI.

Influence of Prior MI on Circadian Pattern

The influence of prior MI on the circadian pattern of acute MI has been investigated in four studies[15,21–23] (Fig. 10). Hansen et al. observed a significantly nonuniform ($p<0.001$) circadian pattern in the overall group with a second quarter peak.[21] However, in those with a history of a prior MI, the second quarter peak was blunted (28.6%) compared to those without a prior MI (31.4%). This subgroup difference was statistically significant ($p=0.033$).[21]

Hjalmarsen and associates also observed blunting of the surge in risk of MI in the second quarter in those with a history of prior MI.[22] In the study by Hjalmarsen, a significantly nonuniform ($p<0.001$) circadian pattern was observed in the overall group, while the prior MI subgroup lacked circadian periodicity. This difference was of marginal sig-

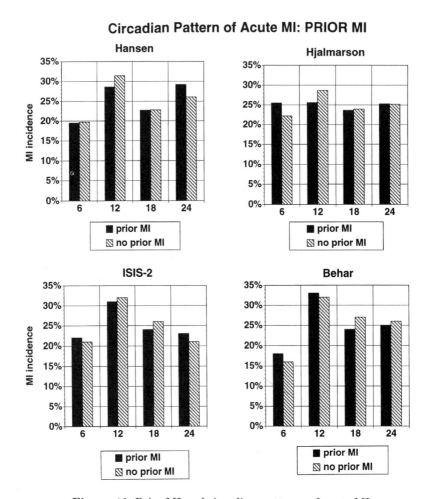

Figure 10: Prior MI and circadian patterns of acute MI.

nificance (p<0.1). Nevertheless, Hansen et al. and Hjalmarsen and associates observed a prior MI to reduce the incidence of MI in the second quarter of the day.[21,22] However, in the ISIS-2 study and in the study by Behar et al., a prior MI was not noted to alter the significant circadian pattern seen in the overall group.[15,23]

Thus, a prior MI has been observed to: (1) abolish circadian periodicity, (2) blunt circadian periodicity without abolishing the pattern, or (3) have no effect. The reasons for the reported differences are not clear. On one hand, those with a prior MI would be more likely to have impaired left ventricular function, suggesting that the demands of waking up and getting out of bed can trigger greater sympathetic activation than that observed in healthy subjects. If this analysis is correct, the history of prior MI should be associated with an increase in the incidence of MI in the second quarter, but this was not observed in any of the prior MI subgroups. Other factors consistent with the blunting of the second quarter seen in some of these studies could include a higher use of medications blunting circadian periodicity due to a prior MI (e.g., beta blockers and aspirin). It is also possible that prior MI patients would have reduced levels of physical activity in the morning by choice or through loss of employment. In this regard, a study by Willich et al. observed septadian periodicity of acute MI onset to be present in workers, but absent in nonworkers.[38] However, it is also important to realize that a circadian pattern can be blunted by either blunting the peak or by having a higher trough, and such a possibility should also be considered for the observed differences in those with a prior MI.

Circadian Pattern in Diabetes

The influence of diabetes on the circadian pattern of acute MI has been reported in six studies[15,16,18,19,21–23] (Fig. 11) The ISIS-2 investigators observed a circadian pattern with a second quarter peak in the overall group that was significantly nonuniform (p<0.001).[23] In the ISIS-2 study, diabetics also demonstrated nonuniformity with a significant (p<0.001) second quarter peak. However, diabetics had a blunted second quarter (28%) vs. nondiabetics (32%) and this difference was significant (p<0.01). The GISSI-2 investigators observed a significant (p<0.001) second quarter peak in the overall group.[18] The circadian pattern of MI in diabetics differed significantly from that of nondiabetics (p<0.01). Diabetics had a higher number of MIs in the first and second quarters but lower risk in the third and fourth quarters compared to nondiabetics. Thus, in the GISSI-2 study, diabetes blunted the magnitude of trough-to-peak transition, although the peak in the second quarter was slightly higher. Hjalmarson et al. observed a significantly nonuniform circadian pattern in the overall group (p<0.001), with a second quarter (28%) peak.[22] Diabetics, however, had a different circadian pattern with equal second (27.5%) and fourth (27.8%) quarter peaks.[22] Tests of independence revealed diabetics to differ from nondiabetics on a marginal basis (p<0.1).[22]

In contrast to these three studies, other investigations have not observed diabetes to alter the circadian pattern of MI. Hansen et al. observed a significantly nonuniform circadian pattern (p<0.001), with a prominent second quarter peak (28.2%) and a smaller fourth quarter peak (26.9%).[21] This circadian pattern was also significant in diabetics

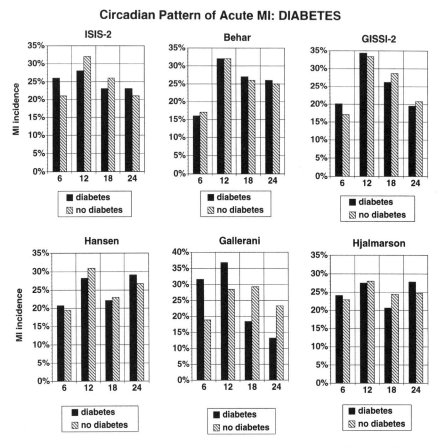

Figure 11: The effect of diabetes on circadian patterns of acute MI.

(p<0.01), but almost equal peaks were noted in the second (28.2%) and fourth (29.1%) quarters. However, the tests of independence did not find a significant difference between diabetics and nondiabetics with regard to their circadian patterns. Behar et al. observed a significant (p<0.01) circadian pattern with a second quarter peak (32%) and a first quarter trough (16%).[15] Subgroup analyses in the study by Behar et al. revealed that the second quarter peak persisted in diabetics (present 32%, absent 32%).[15] Gallerani et al. found no significant circadian pattern in an overall group of 424 acute MI patients as well as in those with diabetes.[16]

Although more work is needed, the available data demonstrated that diabetes has been observed to blunt the circadian pattern of MI in some studies. It is therefore worth considering possible mechanisms. First, diabetic patients may have impaired sympathetic activation, due to autonomic dysfunction, which would be anticipated to blunt triggering of MIs in the second quarter. Secondly, the sequela of this disease may lead to overall decreased levels of physical activity. For example, the associated peripheral vascular disease causing claudication, foot ulcers, and amputations would limit physical activity in the morning in diabetics. The lack of uniformity in the observed circadian pattern of MI in diabetic patients may predominantly be related to the degree of autonomic nervous system dysfunction induced by diabetes. It was recently described that the circadian pattern of myocardial ischemia with peak occurrence in the second quarter of the day is related to the degree of autonomic nervous system dysfunction.[54] In that study by Waxman et al., diabetics lacking autonomic nervous system dysfunction demonstrated a significant second quarter peak for ischemic episodes, while diabetic patients demonstrating autonomic nervous system dysfunction did not have the circadian periodicity.[54]

Influence of Hypertension on Circadian Pattern

The influence of hypertension on the circadian pattern of acute MI has been investigated in four studies, all of which showed a significantly nonuniform circadian periodicity in the overall group.[15,18,21,22] Hansen et al. and the GISSI-2 group observed identical circadian patterns for hypertensive and nonhypertensive subgroups.[18,21] In contrast, Behar et al. observed hypertensives to have reduced second quarter peak and an increased fourth quarter peak compared to nonhypertensives.[15] The difference, however, was not tested for significance. Hjalmarson et al.[22] observed a significant circadian pattern in hypertensives, but it was not significantly different from the nonhypertensive subgroup.[22] Thus, the available data would suggest that hypertension does not modify the circadian pattern of MI.

Influence of Congestive Heart Failure

Hjalmarson et al. observed a significantly nonuniform (p<0.001) circadian pattern in the overall group with a second quarter peak, while the subgroup with a history of prior congestive heart failure demonstrated a fourth quarter peak.[22] However, the differences between the subgroups with and without prior congestive heart failure were not significant. Therefore, at present, no conclusions can be drawn with regard to the influence of prior congestive heart failure on circadian patterns of MIs due to the limited data currently available in this area.

Conclusions

Acute MI demonstrates circadian periodicity, with the most typical pattern consisting of a nadir during hours normally associated with sleep followed by a rapid increase to a peak after arising. The first quarter trough and second quarter peak are followed by intermediate values in the third and fourth quarters of the day. The process of awakening and arising with resultant sympathetic activation appears to play a major role in determining the circadian periodicity of MI. Sympathetic activation increases cardiac work, which can lead to intracoronary plaque rupture. In the morning after arising, there is also a thrombogenic milieu due to increased platelet aggregability and a lowered intrinsic fibrinolytic activity.

Other factors also may be involved in modifying the circadian pattern of MI, such as the presence of triggers and/or protective factors. The use of cardiovascular medications and the influence of age, gender, or associated medical conditions can also alter the circadian manifestations of acute MI. Studies on the circadian effects of cardiovascular medications have revealed that beta blockers can blunt circadian periodicity, an effect consistent with the role of sympathetic activation in triggering MI. Furthermore, although it has been thus far demonstrated in only one study, blunting of circadian periodicity of MIs by aspirin suggests a role for platelet aggregation as a causative factor in the triggering of MI in the morning.

The circadian pattern of MI has been shown to be modified by smoking, age, non-Q-wave MI, diabetes, prior MI, and peripheral vascular disease, but not by gender, hypertension, or prior congestive heart failure. The blunting of circadian periodicity in those with peripheral vascular disease, diabetes, or a prior MI may relate to decreased physical activity levels in those individuals and a lesser degree of sympathetic activation in the morning after arising. Whatever the precise reasons might be for these alterations, it is evident that a variety of factors influence the circadian periodicity of the onset of MI. Although clearly more work is needed in this area, based on the available data, it seems reasonable to conclude that for most individuals the risk of MI onset appears to be greater in the first few hours after waking up, getting out of bed, and resuming daily activities. As discussed later in Chapters 2 and 3, a variety of factors are responsible for the circadian pattern of MIs. It is reasonable to postulate that for any given patient, the critical balance (or lack of it) between various triggers and the protective factors is what determines the risk of MI at any given point in time. Future studies concentrating in these areas should provide further insight regarding the mechanistic aspects of the circadian periodicity so that appropriate therapeutic strategies can be planned to prevent the onset of MI and other related CV disorders.

References

1. Clair WK, Wilkinson WE, McCarthy EA, et al: Spontaneous occurrence of symptomatic paroxysmal atrial fibrillation and paroxysmal supraventricular tachycardia in untreated patients. Am J Cardiol 1993; 87:1114.
2. Rostagno C, Taddei T, Paladini B, et al: The onset of symptomatic atrial fibrillation and paroxysmal supraventricular tachycardia is characterized by different circadian rhythms. Am J Cardiol 1993; 71:453–455.
3. Canada WB, Woodward W, Lee G, et al: Circadian rhythm of hourly ventricular arrhythmia frequency in man. Angiology 1983; 34:274–282.
4. Raeder EA, Hohnloser SH, Graboys TB, et al: Spontaneous variability and circadian distribution of ectopic activity in patients with malignant ventricular arrhythmia. J Am Coll Cardiol 1988; 12:656–661.
5. Lucente M, Rebuzzi AG, Lanza GA, et al: Circadian variation of ventricular tachycardia in acute myocardial infarction. Am J Cardiol 1988; 62:670–674.
6. Twidale N, Taylor S, Heddle WF, et al: Morning increase in the time of onset of sustained ventricular tachycardia. Am J Cardiol 1989; 64:1204–1206.
7. Muller JE, Ludmer PL, Willich SN, et al: Circadian variation in the frequency of sudden cardiac death. Circulation 1987; 75:131–138.
8. Willich SN, Levy D, Rocco MB, et al: Circadian variation in the incidence of sudden cardiac death in the Framingham Heart Study population. Am J Cardiol 1987; 60:801–806.
9. Tsementzis SA, Gill JS, Hitchcock ER, et al: Diurnal variation of and activity during the onset of stroke. Neurosurgery 1987; 17:901–904.
10. Marler JR, Price TR, Clark GL, et al: Morning increase in onset of ischemic stroke. Stroke 1989; 20:473–476.
11. Hausmann D, Nikutta P, Trappe HJ, et al: Circadian distribution of the characteristics of ischemic episodes in patients with stable coronary artery disease. Am J Cardiol 1990; 66:668–672.
12. Nademanee K, Intarachot V, Josephson MA, et al: Circadian variation in occurrence of transient overt and silent myocardial ischemia in chronic stable angina and comparison with Prinzmetal angina in men. Am J Cardiol 1987; 60:494–498.
13. Behar S, Reicher-Reiss H, Goldbourt U, et al: Circadian variation in pain onset in unstable angina pectoris. Am J Cardiol 1991; 67:91–93.
14. Beamer AD, Lee TH, Cook EF, et al: Diagnostic implications for myocardial ischemia of the circadian variation of the onset of chest pain. Am J Cardiol 1987; 60:998–1002.
15. Behar S, Halabi M, Reicher-Reiss H, et al: Circadian variation and possible external triggers of onset of myocardial infarction. SPRINT Study Group. Am J Med 1993; 94:395–400.

16. Gallerani M, Manfredini R, Ricci L, et al: Circadian variation in the onset of acute myocardial infarction: lack of an effect due to age and sex. J Int Med Res 1993; 21:158–160.

17. Ganelina IE, Borisova IY: Circadian rhythm of working capacity, sympathicoadrenal activity, and myocardial infarction. Human Physiol 1983; 9:113–120.

18. Gnecchi-Ruscone T, Piccaluga E, Guzzetti S, et al: Morning and Monday: critical periods for the onset of acute myocardial infarction. The GISSI 2 Study experience. Eur Heart J 1994; 15:882–887.

19. Gnecchi-Ruscone T, Guzzetti S, Piccaluga E, et al: Sleeping hours: a relatively protected period for impending myocardial infarction. Int J Cardiol 1987; 16:161–167.

20. Goldberg RJ, Brady P, Muller JE, et al: Time of onset of symptoms of acute myocardial infarction. Am J Cardiol 1990; 66:140–144.

21. Hansen O, Johansson BW, Gullberg B: Circadian distribution of onset of acute myocardial infarction in subgroups from analysis of 10,791 patients treated in a single center. Am J Cardiol 1992; 69:1003–1008.

22. Hjalmarson A, Gilpin EA, Nicod P, et al: Differing circadian patterns of symptom onset in subgroups of patients with acute myocardial infarction. Circulation 1989; 80:267–275.

23. ISIS-2 (Second International Study of Infarct Survival) Collaborative Group: Morning peak in the incidence of myocardial infarction: experience in the ISIS-2 trial. Eur Heart J 1992; 13:594–598.

24. Johannsson BW: Myocardial infarction in Malmo 1960–1968. Acta Med Scand 1972; 91:505–515.

25. Kleiman NS, Schechtman KB, Young PM, et al: Lack of diurnal variation in the onset of non-Q wave infarction. Circulation 1990; 81:548–555.

26. Kumar PD, Sahasranam KV: Circadian variation in the onset of pain of acute myocardial infarction in Indian patients. J Assoc Physicians India 1992; 40:626–627.

27. Marchant B, Ranjadayalan K, Stevenson R, et al: Circadian and seasonal factors in the pathogenesis of acute myocardial infarction: the influence of environmental temperature. Br Heart J 1993; 69:385–387.

28. Master AM: The role of effort and occupation (including physicians) in coronary occlusions. JAMA 1960; 174:942–948.

29. Muller JE, Stone PH, Turi ZG, et al: Circadian variation in the frequency of onset of acute myocardial infarction. N Engl J Med 1985; 313:1315–1322.

30. Myers A, Dewar HA: Circumstances attending 100 sudden deaths from coronary artery disease with coroner's necropsies. Br Heart J 1975; 37:1133–1143.

31. Pell S, D'Alonzo CA: Acute myocardial infarction in a large industrial population: report of a 6-year study of 1,356 cases. J Am Med Assoc 1963; 185:831–838.

32. Ridker PM, Manson JE, Buring JE, et al: Circadian variation of acute myocardial infarction and the effect of low-dose aspirin in a randomized trial of physicians. Circulation 1990; 82:897–902.

33. Thompson DR, Blanford RL, Sutton TW, et al: Time of onset of chest pain in acute myocardial infarction. Int J Cardiol 1985; 7:139–146.

34. Thompson DR, Sutton TW, Jowett NI, et al: Circadian variation in the fre-

quency of onset of chest pain in acute myocardial infarction: time of onset of chest pain in acute myocardial infarction. Br Heart J 1991; 65:177–178.

35. Myocardial infarction community registers: results of a WHO collaborative study coordinated by the regional office for Europe. World Health Organization Regional Office for Europe Myocardial Infarction Community Registers Copenhagen: Public Health in Europe No. 5, 1976.

36. Willich SN, Linderer T, Wegscheider K, et al: Increased morning incidence of myocardial infarction in the ISAM Study: absence with prior beta-adrenergic blockade. Circulation 1989; 80:853–858.

37. Willich SN, Lowel H, Lewis M, et al: Association of wake time and the onset of myocardial infarction: Triggers and Mechanisms of Myocardial Infarction (TRIMM) pilot study. TRIMM Study Group. Circulation 1991; 84(Suppl):VI62–67.

38. Willich SN, Lowel H, Lewis M, et al: Weekly variation of acute myocardial infarction. Increased Monday risk in the working population. Circulation 1994; 90:87–93.

39. Woods KL, Fletcher S, Jagger C: Modification of the circadian rhythm of onset of acute myocardial infarction by long-term antianginal treatment. Br Heart J 1992; 68:458–461.

40. Zornosa J, Smith M, Little W: Effect of activity on circadian variation at time of acute myocardial infarction. Am J Cardiol 1992; 69:1089–1090.

41. Millar-Craig MW, Bishop CN, Raftery EB: Circadian variation of blood pressure. Lancet 1978; 1(4):795–797.

42. Gould BA, Hornung RS, Mann S, et al: Slow channel inhibitors verapamil and nifedipine in the management of hypertension. J Cardiovasc Pharmacol 1982; 4:S369–373.

43. Chau NP, Mallion JM, de Gaudemaris R, et al: Twenty-four hour ambulatory blood pressure in shift workers. Circulation 1989; 80:341–347.

44. Dewood MA, Stifter WF, Simpson CS, et al: Coronary arteriographic findings soon after non-Q-wave myocardial infarction. N Engl J Med 1986; 315:417–422.

45. Dewood MA, Spores J, Notske R, et al: Prevalence of total coronary occlusion during the early hours of transmural myocardial infarction. N Engl J Med 1980; 303:897–902.

46. Turton MB, Deegan T: Circadian variations of plasma catecholamine, cortisol and immunoreactive insulin concentrations in supine subjects. Clin Chim Acta 1974; 55(3):389–397.

47. Tofler GH, Brezinski D, Schafer AI, et al: Concurrent morning increase in platelet aggregability and the risk of myocardial infarction and sudden cardiac death. N Engl J Med 1987; 316:1514–1518.

48. Peters RW, Zoble RG, Liebson PR, et al: Identification of a secondary peak in myocardial infarction onset 11–12 hours after awakening: the Cardiac Arrhythmia Suppression Trial (CAST) experience. JACC 1993; 22:998–1003.

49. Andreotti F, Davies GJ, Hackett DR, et al: Major circadian fluctuations in fibrinolytic factors and possible relevance to time of onset of myocardial infarction, sudden cardiac death and stroke. Am J Cardiol 1988; 62(9):635–637.

50. Petralito A, Mangiafico RA, Gibiino S, et al: Daily modifications of plasma fibrinogen platelets aggregation, Howell's time, PTT, TT, and antithrombin II in normal subjects and in patients with vascular disease. Chronobiologia 1982; 9(2):195–201.

51. Steering Committee of the Physicians' Health Study Research Group: Final report on the aspirin component of the ongoing Physician's Health Study. N Engl J Med 1989; 321:129–135.
52. Lewis HD, Davis JW, Archibald DG, et al: Protective effects of aspirin against acute myocardial infarction and death in men with unstable angina. N Engl J Med 1983; 309:396–403.
53. Willich SN, Pohjola-Sintonen S, Bhatia SJ, et al: Suppression of silent ischemia by metoprolol without alteration of morning increase of platelet aggregability in patients with stable coronary artery disease. Circulation 1989; 79(3):557–565.
54. Waxman S, Zarich R, Freeman R, et al: Absence of a morning peak of ambulatory ischemia in diabetic patients is related to autonomic dysfunction. JACC 1994; 23:319A.

2

Pathophysiological Basis of Cardiovascular Circadian Rhythmicity

Jaime E. Murillo, MD, Geoffrey H. Tofler, MD

Introduction

Until recently, a relatively neglected area of cardiovascular research has been the mechanism of the conversion from stable to unstable coronary artery disease and why some individuals with coronary atherosclerosis remain asymptomatic or have chronic stable angina whereas others with a similar extent of atherosclerosis develop acute myocardial infarction or sudden cardiac death. Several key advances in understanding of pathophysiology now provide the opportunity to develop improved treatment and prevention strategies. First, the importance of plaque rupture and thrombosis in onset of myocardial infarction (MI)[1,2] has become clear. Second, Little and others[3] have demonstrated that plaques that lead to acute occlusion often have only a mild degree of stenosis, such that "obstructive" plaques that cause angina are not necessarily the same as those that are "vulnerable" to plaque rupture and thrombosis. A third advance that has stimulated the field[4] has been the recognition that time of onset of cardiac events is not random but instead shows a circadian pattern of onset. In the decade since the 1985 observation by Muller et al.[5] that the frequency of onset of MI peaks at 9 AM, numerous publications[6–9] have supported this observation not

From: Deedwania PC (ed): *Circadian Rhythms of Cardiovascular Disorders.* ©Futura Publishing Co., Inc., Armonk, NY, 1997.

only for MI, but also for sudden cardiac death, transient myocardial ischemia, and stroke.

Refinement of these epidemiological observations has led, first, to the conclusion that the morning peak in disease onset is due in part to the physical and mental stressors associated with morning awakening and activity and, second,[10] that stressors such as heavy physical activity and anger can trigger acute cardiovascular events.

Myocardial Infarction

Investigations of the cause of the morning increase in MI are aided by recent advances in understanding the pathological basis of this disorder.

The importance of plaque rupture and thrombosis in onset of MI has led investigators to identify several factors that, in the presence of a vulnerable plaque, may increase the risk of disease onset in the morning (Fig. 1). The surge in arterial blood pressure in the morning[11] could induce the rupture of an atherosclerotic plaque. Increases in arterial pressure and contractility[12,13] may elevate shear stress forces at a plaque, particularly at the site of transition between lipid-rich pool and relatively normal endothelium. These increased shear forces may predispose to plaque injury. Increased shear stress in the morning may also be produced by the heart rate surge that occurs on morning awakening and onset of activities. In human and animal studies, arterial tone[14] has been demonstrated to be increased in the morning hours. This increased tone could worsen flow reduction at the site of a fixed stenosis and result in increased shear stress predisposing to plaque rupture. Fujita and Franklin[15] reported that coronary blood flow in dogs was substantially lower in the morning compared to the afternoon under resting hemodynamic conditions.

Several hemostatic factors show morning increases in activity that predispose to a prothrombotic state. Platelet aggregability[16,17] is increased in the morning, following assumption of the upright posture. Morning increases in viscosity[18] have also been reported. Fibrinolytic activity exhibits a marked diurnal fluctuation[19,20] characterized by a nadir in the morning and a peak activity between 5 and 8 PM. This circadian variation is due primarily to the fluctuation in plasminogen activator inhibitor (PAI-1) levels that peak in the early morning hours. Although the mechanism of this variation has not been fully explained, its amplitude is reduced[20,21] when subjects remain resting in bed. The relative hypercoagulability in the morning could increase the likelihood that a recently formed small thrombus overlying a plaque rupture would propagate and lead to a total occlusion of the vessel.

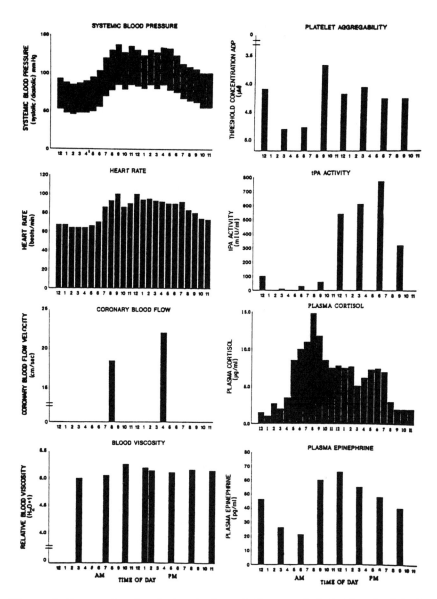

Figure 1: Bar graphs of variation during a 24-hour period of eight physiological processes possibly contributing to the increased morning frequency of MI. Systemic blood pressure and heart rate[11] measured intra-arterially in five normotensive ambulant subjects; coronary blood flow velocity[15] measured by Doppler ultrasonic flow probe in 21 dogs; whole-blood viscosity[18] measured by Ostwald-capillary-viscometer in eight normal male volunteers; platelet aggregability[16] measured by in vitro platelet aggregometry in 15 normal male volunteers; tissue-type plasminogen activity (t-PA)[72] measured by spectrophotometric assay in six normal volunteers (four male, two female); plasma cortisol[22] measured by competitive protein binding method in six normal male volunteers; plasma epinephrine[16] measured by a radioisotope method in 15 normal male volunteers. See text for discussion. (Adapted with permission.[4])

Morning peaks in plasma catecholamines[16] and cortisol[22] may also contribute to the morning peak of disease onset. Changes in plasma epinephrine and norepinephrine are largely based on the activity/rest cycle whereas cortisol has a true endogenous circadian pattern. Although serum cortisol levels peak earlier in the morning than catecholamines, and fall during the period of increased disease onset, the levels remain increased above basal at the time of peak frequency of infarction. The increased morning cortisol level may sensitize the coronary vessels to the vasoconstrictor effects of catecholamines, which surge after assumption of the upright posture and onset of activity. This possibility is supported by data from Sudhir,[23] indicating that hydrocortisone increases forearm vascular responsiveness to cold-pressor stimulation and to norepinephrine. Thus, increased catecholamine levels in the morning, when cortisol levels are high, may have a more pronounced effect on vascular tone than if a catecholamine surge occurred at a time of low cortisol levels.

Acute onset of a myocardial infarction may thus be triggered by a simultaneous increase in activity of a number of factors, benign when appearing alone but potentially pathological in combination.

An important component in the process leading to thrombosis is the presence of a vulnerable atherosclerotic plaque. This vulnerability can be understood as a dynamic and potentially reversible process. Thus, onset might begin when physical or mental stress triggers a hemodynamic change sufficient to rupture a vulnerable plaque. However, if such a trigger does not occur during vulnerability, the plaque may change and become nonvulnerable.

Sudden Cardiac Death

Although sudden cardiac death is ultimately an electrical event, approximately one-third of the subjects have a fresh, occlusive thrombus at autopsy examination. Davies and Thomas[1] have presented data indicating that even the two-thirds of cases of sudden cardiac death without occlusive thrombus frequently have a nonocclusive thrombus that may lead to temporary occlusion or distal embolization. For the large number of cases of sudden cardiac death in which plaque injury and thrombosis play a role, the physiological processes described for MI apply. In the absence of thrombosis, a primary ventricular tachyarrhythmia in the absence of infarction is also more likely to occur in the morning hours. Most studies of ventricular ectopy indicate a prominent peak during daytime hours and a trough during the night. In 164 ambulatory patients studied over 3 consecutive days using 24-hour Holter monitoring, Canada et al.[24] showed that a circadian pattern of ventricular premature beats with a morning peak was consistently present for each day of observation. Twidale et al.[25] observed that the peak incidence of sustained, symptomatic ventricular tachycardia episodes in 68 patients occurred between 10 AM and noon. The recent availability of implantable cardioverter defibrillators that record the time of ventricular tachyarrhythmias requiring either pacing or shock therapy provides an opportunity to clarify the timing of ventricular tachyarrhythmias predisposing to sudden cardiac death. Our group and others[26,27] have found that a higher proportion of ventricular tachyarrhythmias begin in the late morning compared with other times of the day.

In contrast to ventricular tachyarrhythmias, asystolic arrest does not exhibit a morning peak in frequency. Investigations into heart rate variability (HRV) provide insight into the mechanism of the circadian variation in arrhythmia. Reduced heart rate variability, as measured by the standard deviation of RR intervals over 24 hours, indicates a depressed vagal to sympathetic balance and is associated with increased mortality after myocardial infarction. Heart rate variability[28,29] also exhibits a circadian rhythm with the lowest values in the late morning. This circadian pattern of HRV, which is observed among normal subjects[28] as well as among those with coronary artery disease, is blunted in survivors of cardiac arrest. The increased adrenergic tone and the withdrawal of vagal tone in the morning[30,31] may create an interval with a lower threshold for malignant arrhythmias.

Transient Myocardial Ischemia

Since transient myocardial ischemia is associated with an increased risk for MI and sudden cardiac death, study of the timing of transient ischemia provides insight into the mechanism of disease onset. The use of 24-hour ambulatory Holter monitoring has enabled the timing of transient ischemia to be carefully and precisely studied. A complete coverage of symptomatic as well as asymptomatic episodes of myocardial ischemia is possible, thereby eliminating bias resulting from unobserved periods. These data are detailed in other chapters; however, a circadian pattern, characterized by a morning peak in frequency, has been reported. Rocco et al., adjusting for wake-time, showed that the increase in frequency of ischemia[7] occurs in the first 2 hours after awakening and onset of morning activity. More recently, Parker et al.[32] convincingly demonstrated that the morning increase in transient ischemia occurred not when subjects remained resting in bed but only when they began activity.

An important mechanism for the morning peak in ischemia is the increased myocardial oxygen consumption due to the simultaneous rise in heart rate, blood pressure, and catecholamines during the morning.

Ischemia may also be produced by a decrease in myocardial oxygen supply[7,33,34] (due to transient vasoconstriction or platelet aggregation leading to decreased coronary blood flow). Exercise performance and ischemic threshold in patients with stable angina have been reported[35] to be lower early in the morning than in the afternoon, paralleling changes in vascular resistance.

Platelet activation and thrombotic mechanisms do not appear to play a significant role in stable exertional angina. This conclusion is based on the inability of platelet inhibitors such as aspirin and ticlopidine[36] to reduce episodes of transient ischemia. Furthermore, metoprolol[37] reduced the frequency and duration of ischemic episodes without affecting the morning surge in platelet aggregability. However, platelet activation plays a major role in patients with unstable angina. Unstable angina[38] has also been shown to peak in frequency in the morning hours.

A circadian pattern has also been described for variant angina associated with ST elevation,[39–41] although episodes peak earlier in the morning at 4 AM, often waking the patients. Variant anginal attacks[42] have also been shown to be more readily provoked by exercise in the morning than in the afternoon. Coronary spasm plays an important role in episodes of variant angina, often occurring at the site of a major coronary stenosis.

Stroke

A circadian pattern of both ischemic and hemorrhagic stroke[9,43] has been demonstrated, with a morning peak similar to that of other cardiovascular diseases. The mechanism of this pattern has not been well investigated; however, many of the processes previously described could contribute. These include the increased hemodynamic forces acting on a atherosclerotic plaque, a relatively prothrombotic state, and increased arrhythmogenesis.

Triggering of Cardiovascular Disease

Although attention has focused on the morning as the time of peak incidence of disease onset, similar physiological processes probably trigger disease onset at other times of the day. The peak morning incidence of sudden cardiac death and MI probably results from the synchronization of the population for triggers in the morning, whereas a secondary evening peak in MI onset observed in the MILIS data[5] may

Table 1

Stressors that May Trigger the Onset of Cardiovascular Events

Possible Triggers	Systemic Arterial Pressure	Heart Rate	Coronary Vascular Resistance or Narrowing*	Plasma Catecholamines	Platelet Activity	Fibrinolytic Activity
1. Exercise	↑	↑	↑	↑	↑↔	↑
2. Assumption of upright posture	↓↔	↑	?	↑	↑	↑
3. Handgrip	↑	↑	↑	↑	?	?
4. Cold exposure	↑	↑	↑↔	↑	↑↔	↑
5. Cigarette smoking	↑	↑	↑	↑	↑↔	↑ acute ↓ chronic
6. Mental stress	↑	↑	↑	↑	↑	↑

↑ = increased; ↓ = decreased; ↔ = no change; * = response observed with coronary atherosclerosis.
Adapted with permission.[4]

result from synchronization of the population for an additional trigger such as the evening meal. For other periods of the day, exposure of the population to potential triggers is random, and no other prominent peaks of incidence are observed.

Uncertainty exists as to the degree to which the 24-hour disease periodicity results from a true, endogenous circadian rhythm versus the daily rest-activity cycle. It is likely that the rest-activity cycle is a major determinant of disease onset, since adjustment for time of awakening increases the relative risk of the morning occurrence of sudden cardiac death,[44] nonfatal infarction,[8] and transient ischemia.[7] However, this adjustment could also align the population for their endogenous circadian rhythms. The morning physiological changes also vary in their circadian pattern. Platelet aggregability increase, for example, is abolished if the subject[17] remains in bed, whereas cortisol secretion[22] is not affected by the time of onset of activity. The circadian pattern of fibrinolytic activity persists if subjects remain in bed[19]; however, the amplitude of variation is reduced.

Many of the physiological changes that occur in the morning also occur in response to stressors that may trigger the onset of cardiovascular events (see Table 1).

Physical Exertion

Transition from a sleeping state to the onset of physical exertion is an important component of the morning period. Controlled studies have now firmly identified physical exertion as a trigger of nonfatal MI, sudden cardiac death, and transient ischemia. The relative risk of nonfatal infarction[10] was increased 5.9-fold in the hour following heavy exertion in the ONSET Study. Moreover, the risk was increased to 107-fold in sedentary individuals, but those who regularly exercised had only a two-fold increase in risk. Several physiological responses to exercise may increase the risk of cardiovascular disease. Adrenergic stimulation is associated with a surge in heart rate, blood pressure, and contractility. While vasodilatation occurs in normal coronary arteries[45] in response to dynamic exercise, paradoxic vasoconstriction may occur in coronary arteries complicated by atherosclerosis. Atherosclerotic arteries show an exaggerated vasomotor sensitivity to normally occurring vasoconstrictor substances,[46,47] such as noradrenaline and serotonin, and a paradoxic vasoconstrictor response[48,49] to acetylcholine. In the presence of atherosclerosis, handgrip exercise[50] has also been associated with vasoconstriction.

Fibrinolytic activity normally increases with exercise; however, this beneficial response is reduced in patients with atherosclerosis. The increase in tissue plasminogen activator (t-PA) activity[51] is also less in sedentary individuals than in those who regularly exercise. Platelet activation may also be produced by physical stress, more so in individuals with coronary artery disease than in healthy subjects.

Mental Stress

While numerous anecdotal cases of mental stress acutely triggering infarction and sudden cardiac death have been reported, controlled data have been limited. However, several studies of patients with coronary artery disease have found that mental stress increases the risk of transient ischemia. In addition to the well-documented increase in blood pressure and heart rate during mental stress that would increase myocardial oxygen demand, Deanfield et al.[52] found that mental arithmetic was associated with a reduction in regional coronary artery blood flow in patients with stable angina. Therefore, ischemia that follows mental stress might result in part from a primary decrease in coronary blood flow. In dogs with a partially obstructed coronary artery, Verrier et al.[53] showed that coronary vascular resistance increased and blood flow decreased during episodes of anger and that a further 35% reduction in flow occurred 2–4 minutes after the episode.

Emotional stress may also produce an autonomic imbalance characterized by sympathetic predominance. In a spectral analysis of the HRV obtained during laboratory mental stress, Pagani et al.[30] showed an increase in the low-frequency (i.e., sympathetic tone) and decrease in the high-frequency (i.e., the vagal tone) power spectra. The prolonged QT syndrome[54] illustrates the association of mental stress, adrenergic stimulation, arrhythmia, and SCD. β-adrenergic blockade may be of particular benefit in reducing events triggered by mental stress.

Cold

The winter months have been associated with a 60% increase in cardiovascular mortality compared to summer months. Anderson[56] estimated that a 4.4°C decline in temperature within 24 hours, during the winter, was associated with a 16% increase in the number of sudden coronary deaths among men over 65 years of age, and similarly an 8.3°C decline in temperature was associated with a 25% increase in sudden coronary death. They also reported an increased incidence of MI associated with snowfalls. An increase in risk factors such as arterial blood pressure[57] and fibrinogen occurs during the winter months. Acutely, the cold-pressor test[58] has been shown to increase vascular resistance.

Smoking

Smoking has physiological effects that acutely increase the risk of cardiovascular disease, including vasoconstriction, platelet activation, and blood pressure and heart rate surges. Somewhat paradoxically, smoking may acutely increase t-PA activity; however, smokers have chronically lower levels of fibrinolytic potential than nonsmokers.

The identification of stressors as triggers of cardiovascular disease has led to the concept of the acute risk factor, defined as a transient physiological change, such as a surge in arterial pressure and increase in coagulability or vasoconstriction, that may trigger disease onset. The manner in which different individuals respond to stressors may determine the extent of their transient increase in risk following the stress. Total risk at a particular time can be considered to be a sum of chronic risk factors and the level of acute risk (Fig. 2).

Pharmacological Modification of the Circadian Pattern of Cardiovascular Disease

Evaluation of the effect of medications on the temporal pattern of cardiovascular disease onset may provide a better insight into the

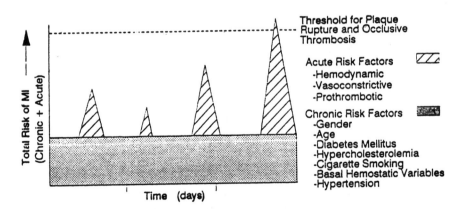

Figure 2: This figure represents the hypothesis that the risk of an acute event is a dynamic process determined by chronic risk factors that change slowly with time and acute risk factors that may be rapidly generated by external stresses. In this figure, the spikes indicate transient increases in risk resulting from a stress such as anger or physical exertion. These spikes rise from a baseline of chronic risk.

mechanism of the disease onset and also aid in the design of improved preventive strategies.

β-Adrenergic Blockers

It has been known for some time that β-blockers[59,60] reduce the incidence of recurrent MI and sudden cardiac death following MI. It has also been reported[61] that these agents are capable of providing primary prevention against MI, although this finding has not been proven. In the MILIS[5] study, patients taking β-blocking agents prior to their myocardial infarction had no morning peak in onset of acute MI. This finding was confirmed in the ISAM and TIMI II[62,63] databases. Patients who were taking calcium channel blockers prior to their MI exhibited a typical morning increase. A retrospective analysis[60,55] of the time of day of out-of-hospital sudden cardiac death in the β-Blocker Heart Attack Trial, a multicenter study involving 3,857 patients after MI randomized to either placebo or propranolol, showed that the major benefit of β-blockade occurred during the morning hours. During this period, there was a 44% reduction in the incidence of sudden cardiac death in the β-blocker-treated group versus a reduction of just 18% at other times of the day. An additional beneficial effect of β-blockers[64] is the morning reduction of silent ischemia episodes that also have a morning peak of incidence in the nontreated group.

Despite extensive study, the mechanism by which β-blockers exert their protective effect has not been identified. It has been particularly difficult to explain the ability of a β-blocker to prevent MI, which is generally caused by coronary thrombosis, since most β-adrenergic blocking agents[65] have only minimal antiplatelet activity. It is possible that β-blockers exert their protective effect by blunting the morning increase in sympathetic activity. Such blunting would minimize increases in myocardial contractility, arterial pressure, and heart rate, which might increase wall stress[66] and the likelihood of plaque disruption and consequent thrombosis.

Aspirin and Other
Antiplatelet Agents

The benefit of aspirin[67] has been shown for both primary and secondary prevention of MI. In the aspirin-treated group[68] of the Physician's Health Study, the morning peak in MI was selectively attenuated (a 59.3% reduction in the incidence of infarction during the morning, compared to a 34.1% reduction for the remaining hours of the day) (Fig. 3).

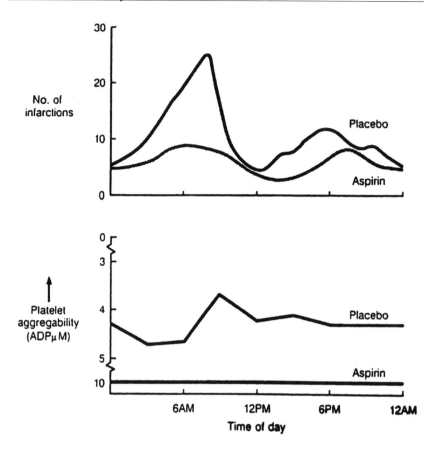

Figure 3: Prevention of morning myocardial infarction by preventing platelet surge. Proposed contribution of morning platelet surge to morning myocardial infarction and its prevention by aspirin. Illustration combines (top) the ability of aspirin to selectively reduce the morning peak in frequency of myocardial infarction (adapted from Ridker et al.[68]) and (bottom) the ability of aspirin to abolish the morning increase in platelet aggregability. ADP = adenosine diphosphate.

This differential effect may reflect the blunting of a potential triggering mechanism[16]—the morning surge in platelet activity.

Safety of Exercise in the Morning

Although the circadian data have led some investigators[69] to question the safety of morning exercise, there is currently no epidemiological data supporting any added risk, especially when one considers the long-term beneficial effects of exercise. Some studies (but not others) have found differences in the magnitude of physiological responses to exercise. For example, Rosing et al.[20] found that fibrinolytic activity increased to a greater extent following afternoon than morning exercise, whereas Jimenez et al.[70] found no difference in fibrinolytic or hemodynamic response to isometric exercise when performed in the evening versus the morning. Several investigators have reported that physical exertion is more likely to produce angina in the morning than in the afternoon. In patients with stable angina, Starling and coworkers[71] reported a longer exercise duration in the afternoon test when compared to the morning.

Significance

The primary immediate value of recognition of the circadian variation of acute onset of MI is the emphasis that can be placed on pharmacological protection during the morning hours for patients already receiving anti-ischemic therapy. Although no data are available to support the hypothesis, it seems reasonable that long-acting anti-ischemic agents would have an advantage over short-acting agents in providing protection against MI in the morning when the effects of short-acting agents taken the night before may begin to attenuate. Of more long-term significance is the insight that these data provide into the mechanism of onset of acute cardiovascular disease.

The risk of cardiovascular disease is not static but instead varies throughout the day and probably throughout the year as well. Disease onset requires a combination of factors, in particular, a vulnerable plaque and an external trigger. Reaching the critical point when acute MI occurs may be a rare event in the lifespan of a person. However, the total sum of these episodes still represents a major burden on our society. A comprehensive understanding of the pathophysiological basis of cardiovascular disease including its circadian variation may lead to new methods of prevention.

Furthermore, the recognition of potential periods of vulnerability during the day may lead to the development of screening methods to identify patients at increased risk of cardiovascular disease and to improved therapeutic approaches and methods for prevention.

References

1. Davies M, Thomas A: Plaque fissuring: the cause of acute myocardial infarction, sudden ischemic death, and crescendo angina. Br Heart J 1985; 53:363.
2. De Wood M, Spores J, Notske R, et al: Prevalence of total coronary occlusion during the early hours of transmural myocardial infarction. N Engl J Med 1980; 303:897.

3. Little W, Constantinescu M, Applegate R, et al: Can coronary angiography predict the site of a subsequent myocardial infarction in patients with mild-to-moderate coronary artery disease? Circulation 1988; 78(5 Pt 1):1157.

4. Muller JE, Tofler GH, Stone PH: Circadian variation and triggers of onset of acute cardiovascular disease. Circulation 1989; 79:733–743.

5. Muller JE, Stone PH, Turi ZG, et al: Circadian variation in the frequency of onset of acute myocardial infarction. N Engl J Med 1985; 313:1315.

6. Willich SN, Levy D, Rocco MB, et al: Circadian variation in the incidence of sudden cardiac death in the Framingham Heart Study population. Am J Cardiol 1987; 60:801.

7. Rocco MB, Barry J, Campbell S, et al: Circadian variation of transient myocardial ischemia in patients with coronary artery disease. Circulation 1987; 75:395.

8. Goldberg R, Brady P, Muller JE, et al: Time of onset of symptoms of acute myocardial infarction. Am J Cardiol 1990; 66:140.

9. Marler JR, Price TR, Clark GL, et al: Morning increase in onset of ischemic stroke. Stroke 1989; 20:473.

10. Mittleman M, Maclure M, Tofler GH, et al: Triggering of acute myocardial infarction by heavy physical exertion. Protection against triggering by regular exercise. N Engl J Med 1993; 329:1677.

11. Millar-Craig MW, Bishop CN, Raftery EB: Circadian variation of blood pressure. Lancet 1978; 1:795.

12. Richardson P, Davies M, Born G: Influence of plaque configuration and stress distribution on fissuring of coronary atherosclerotic plaques. Lancet 1989; 2:941.

13. Lee R, Grodzinsky A, Frank E, et al: Structure-dependent dynamic mechanical behavior of fibrous caps from human atherosclerotic plaques. Circulation 1991; 83:1764.

14. Panza JA, Epstein SE, Quyyumi AA: Circadian variation in vascular tone and its relation to alpha sympathetic vasoconstrictor activity. N Engl J Med 1991; 325:986.

15. Fujita M, Franklin D: Diurnal changes in coronary blood flow in conscious dogs. Circulation 1987; 76:488.

16. Tofler GH, Brezinski D, Schafer A, et al: Concurrent morning increase in platelet aggregability and the risk of myocardial infarction and sudden cardiac death. N Engl J Med 1987; 316:1514.

17. Brezinski DA, Tofler GH, Muller JE, et al: Morning increase in platelet aggregability: association with assumption of the upright posture. Circulation 1988; 78:35.

18. Ehrly AM, Jung G: Circadian rhythm of human blood viscosity. Biorheology 1973; 10:577.

19. Fearnley G, Balmforth G, Fearnley E: Evidence of a diurnal fibrinolytic rhythm: with a simple method of measuring natural fibrinolysis. Clin Sci 1957: 16:645.

20. Rosing DR, Brakman P, Redwood DR, et al: Blood fibrinolytic activity in man: diurnal variation and the response to varying intensities of exercise. Circ Res 1970; 27:171.

21. Tofler GH, Creisler C, Rutherford J, et al: Increased platelet aggregability after arising from sleep (abstract). J Am Coll Cardiol 1986; 7:116A.

22. Weitzman ED, Fukushima D, Nogeire C, et al: Twenty-four hour pattern of

the episodic secretion of cortisol in normal subjects. J Clin Endocr 1971; 33:14.

23. Sudhir K, Jennings GL, Esler MD, et al: Hydrocortisone-induced hypertension in humans: pressor responsiveness and sympathetic function. Hypertension 1989; 13:416.

24. Canada W, Woodward W, Lee G, et al: Circadian rhythm of hourly ventricular arrhythmia frequency in man. Angiology 1983; 34:274.

25. Twidale N, Taylor S, Heddle W, et al: Morning increase in the time of onset of sustained ventricular tachycardia. Am J Cardiol 1989; 64:1204.

26. Gebara O, Mittleman M, Rasmussen C, et al: Morning peak in ventricular arrhythmias detected by time of implantable cardioverter-defibrillator therapy. J Am Coll Cardiol 1994; 204A.

27. Lampert R, Rosenfeld L, Batsford W, et al: Circadian variation of sustained ventricular tachycardia in patients with coronary artery disease and implantable cardioverter-defibrillator. Circulation 1994; 90:241.

28. Furlan R, Guzzetti S, Crivellaro W, et al: Continuous 24-hour assessment of the neural regulation of systemic arterial pressure and RR variabilities in ambulant subjects. Circulation 1990; 81:537.

29. Huikuri HV, Linnaluoto MK, Seppanen T, et al: Circadian rhythm of heart rate variability in survivors of cardiac arrest. Am J Cardiol 1992; 70:610.

30. Pagani M, Mazzuero G, Ferrari A, et al: Sympathovagal interaction during mental stress: a study using spectral analysis of heart rate variability in health control subjects and patients with a prior myocardial infarction. Circulation 1991; 83 (Suppl II):43.

31. Schwartz PJ, La Rovere MT, Vanoli E: Autonomic nervous system and sudden cardiac death: experimental basis and clinical observations for post-myocardial infarction risk stratification. Circulation 1992; 85(Suppl I):77.

32. Parker J, Testa M, Jiménez A, et al: Morning increase in ambulatory ischemia in patients with stable coronary artery disease: importance of physical activity and increased cardiac demand. Circulation 1994; 89:604.

33. Barry J, Selwyn AP, Nabel EG: Frequency of ST-segment depression produced by mental stress in stable angina pectoris from coronary artery disease. Am J Cardiol 1988; 61:989.

34. Singh BN, Nademanee K, Figueras J, et al: Hemodynamic and electrocardiographic correlates of symptomatic and silent myocardial ischemia: pathophysiologic and therapeutic implications. Am J Cardiol 1986; 58:3.

35. Quyyumi AA, Panza JA, Diodati JG, et al: Circadian variation in ischemic threshold: a mechanism underlying the circadian variation in ischemic events. Circulation 1992; 86:22.

36. Khurmi N, Bowles M, Rafferty E: Are anti-platelet drugs of value in the management of patients with chronic stable angina? A study with ticlopidine. Clin Cardiol 1986; 9:493.

37. Willich S, Pohjola-Sintonen S, Bhatia S, et al: Suppression of silent ischemia by metoprolol without altering the morning increase in platelet aggregability in patients with stable coronary artery disease. Circulation 1989; 79:557.

38. Behar S, Reicher-Reiss H, Goldbourt U, et al: Circadian variation in pain onset in unstable angina pectoris. Am J Cardiol 1991; 67:91.

39. Guazzi M, Fiorentini C, Polese A, et al: Continuous electrocardiographic recording in Prinzmetal's variant angina pectoris. Br Heart J 1970; 32:611.

40. Araki H, Koiwaya Y, Nakagaki U, et al: Diurnal distribution of ST segment elevation and related arrhythmias in patients with variant angina: a study of ECG ambulatory monitoring. Circulation 1983; 67:995.

41. Waters D, Miller D, Bouchard A, et al: Circadian variation in variant angina. Am J Cardiol 1984; 54:61.

42. Yasue H, Omote S, Takizawa A, et al: Circadian variation of exercise capacity in patients with Prinzmetal's variant angina: Role of exercise-induced coronary arterial spasm. Circulation 1979; 59:939.

43. Tsementzis SA, Gill JS, Hitchcock ER, et al: Diurnal variation of and activity during the onset of stroke. Neurosurgery 1985; 17:901.

44. Willich SN, Goldberg RJ, Maclure M, et al: Increased onset of sudden cardiac death in the first three hours after awakening. Am J Cardiol 1992; 70:65.

45. Gage JE, Hess OM, Murakami T, et al: Vasoconstriction of stenotic coronary arteries during dynamic exercise in patients with classic angina pectoris: reversibility by nitroglycerine. Circulation 1986; 73:865.

46. Ginsburg R, Bristow MR, Davis K, et al: Quantitative pharmacologic responses of normal and atherosclerotic isolated human epicardial coronary arteries. Circulation 1984; 69:430.

47. Ganz P, Alexander RW: New insights into the cellular mechanisms of vasospasm. Am J Cardiol 1985; 56:11.

48. Furchgott RF, Zawadzki JV: The obligatory role of endothelial cells in the relaxation of arterial smooth muscle by acetylcholine. Nature 1980; 288:373.

49. Ludmer PL, Selwyn AP, Shook TL, et al: Paradoxical vasoconstriction induced by acetylcholine in atherosclerotic coronary arteries. N Engl J Med 1986; 315:1046.

50. Brown BG, Lee AB, Bolson EL, et al: Reflex constriction of significant coronary stenosis as a mechanism contributing to ischemic left ventricular dysfunction during isometric exercise. Circulation 1984; 70:18.

51. Szymanski L, Pate R, Durstine L: Effects of maximal exercise and venous occlusion on fibrinolytic activity in physically active and inactive men. J Appl Physiol 1994; 77:2305.

52. Deanfield JE, Shea M, Kensett M, et al: Silent myocardial ischemia due to mental stress. Lancet 1984; 2:1001.

53. Verrier RL, Hagestad EL, Lown B: Delayed myocardial ischemia induced by anger. Circulation 1987; 75:249.

54. Schwartz PJ, Zaza A, Locati E, et al: Stress and sudden death: the case of the long QT syndrome. Circulation 1991; 83(Suppl II):71.

55. Peters RW, Muller JE, Goldstein S, et al: Propranolol and the morning increase in the frequency of sudden cardiac death (BHAT Study). Am J Cardiol 1989; 63:1518.

56. Anderson T, Rochard C: Cold snaps, snowfall, and sudden death from ischemic heart disease. Can Med Assoc J 1979; 121:1580.

57. Giaconi S, Ghione S, Palombo C, et al: Seasonal influences on blood pressure in high normal to mild hypertensive range. Hypertension 1989; 14:22.

58. Mudge GH, Grossman W, Mills RM, et al: Reflex increase in coronary vascular resistance in patients with ischemic heart disease. N Engl J Med 1976; 295:1333.

59. Norwegian Multicenter Study Group: Timolol induced reduction in mortality and reinfarction in patients surviving acute myocardial infarction. N Engl J Med 1981; 304:801.
60. Beta-blocker Heart Attack Trial Research Group: A randomized trial of propranolol in patients with myocardial infarction: mortality results. JAMA 1982; 247:1707.
61. Wikstrand J, Warnold I, Olsson G, et al. Primary prevention with metoprolol in patients with hypertension: mortality results from the MAPHY study. JAMA 1988; 259:1976.
62. Willich SN, Linderer T, Wegscheider K, et al: Increased morning incidence of myocardial infarction in the ISAM Study: absence with prior beta-adrenergic blockade. Circulation 1989; 80:853.
63. Tofler GH, Muller JE, Stone PH, et al: Modifiers of timing and possible triggers of acute myocardial infarction in the Thrombolysis in Myocardial Infarction Study (TIMI II) population. J Am Coll Cardiol 1992; 20:1045.
64. Mulcahy D, Keegan J, Cunningham D, et al: Circadian variation of total ischaemic burden and its alteration with anti-anginal agents. Lancet 1988; 2:755.
65. Frishman WH: Multifactorial actions of beta-adrenergic blocking drugs in ischemic heart disease: current concepts. Circulation 1983; 67(Suppl I):1.
66. Gertz SD, Roberts WC: Hemodynamic shear force in rupture of coronary arterial atherosclerotic plaques [Editorial]. Am J Cardiol 1990; 66:1368.
67. Hennekens CH, Buring JE, Sandercock P, et al: Aspirin and other antiplatelet agents in the secondary and primary prevention of cardiovascular disease. Circulation 1989; 80:749.
68. Ridker PM, Manson JE, Buring JE, et al: Circadian variation of acute myocardial infarction and the effect of low-dose aspirin in a randomized trial of physicians. Circulation 1990; 82:897.
69. Murray PM, Herrington DM, Pettus CW, et al: Should patients with heart disease exercise in the morning or afternoon? Arch Int Med 1993; 153:833.
70. Jiménez A, Tofler GH, Chen X, et al: Effects of nadolol on hemodynamic and hemostatic responses to potential mental and physical triggers of myocardial infarction in subjects with mild systemic hypertension. Am J Cardiol 1993; 72:47–52.
71. Starling MR, Moody M, Crawford MH, et al: Repeat treadmill exercise testing: variability of results in patients with angina pectoris. Am Heart J 1984; 107:298.
72. Andreotti F, Davies GJ, Hackett DR, et al: Major circadian fluctuations in fibrinolytic factors and possible relevance to time of onset of myocardial infarction, sudden cardiac death, and stroke. Am J Cardiol 1988; 62:635.

3

Coronary Vasomotor Tone and Vascular Reactivity

Arshed A. Quyyumi, MD, David Mulcahy, MD

Introduction

Over the last 15 years, evidence has accumulated to conclusively demonstrate that several transient cardiac events including myocardial ischemia[1-5] and nonsustained arrhythmias,[6] and catastrophic cardiac events such as myocardial infarction (MI),[7,8] sustained ventricular arrhythmias,[9] and sudden cardiac death[10] occur more frequently in the morning hours. These observations provided fuel for the search of possible triggers ranging from local factors to systemic and external phenomena that could be responsible for these events. It was understood that precipitating factors would need to predominate in the morning hours, persist throughout the day, and be less prevalent at night. At the same time it was realized that the underlying pathophysiological mechanisms for acute and subacute coronary phenomena vary tremendously. Myocardial ischemia, for example, is precipitated by an imbalance between myocardial oxygen demand and supply, whereas unstable angina and acute MI are often a result of plaque rupture or disruption with superimposed thrombus formation.[11,12] Sudden cardiac death could follow plaque rupture, thrombosis, platelet embolization, and associated sustained ventricular arrhythmias, or result from pump failure or cardiac rupture after myocardial necrosis.[12] It is thus remarkable that such diverse patho-

From: Deedwania PC (ed): *Circadian Rhythms of Cardiovascular Disorders.* ©Futura Publishing Co., Inc., Armonk, NY, 1997.

physiological phenomena that affect atherosclerotic coronary arteries can cluster at the same time of day. Furthermore, it is unlikely that the same triggers are equally important in precipitating each of these conditions and, therefore, a search for multiple triggers is warranted. In this review, we will examine the contribution of changes in vasomotor tone to precipitation of acute and subacute coronary syndromes. The pathophysiological processes that regulate vasomotor changes and therapeutic options will be discussed in light of recent developments in the understanding of vascular biology.

Transient Myocardial Ischemia

Activity as a Trigger

Several studies have demonstrated a circadian pattern of distribution of myocardial ischemia in patients with coronary artery disease with a surge in the mornings starting around 7 AM, reaching a plateau during the day, perhaps a second evening peak, and with a trough in ischemic activity at night.[1-4] Parker et al. have demonstrated that awakening, arousal, and activities constituted the most important trigger for the morning preponderance of ischemic episodes.[3] By allowing patients to arise later in the day, they demonstrated that the peak in ischemic activity could be delayed to a later time. Rocco et al. had earlier demonstrated a more discrete and striking peak in ischemic activity in the first 2–3 hours after awakening in contrast to a more diffuse increase observed in population studies where the time of awakening varies between individuals.[2] During the first 3 hours after awakening, 24% of all daily ischemic episodes had occurred.

Myocardial Oxygen Demand

With the advent of ambulatory blood pressure and ST segment monitoring, where changes in heart rate and blood pressure could be correlated with ischemic activity, it has been repeatedly shown that the large majority of ischemic episodes are preceded by increases in heart rate[13-17] and blood pressure,[15] suggesting that increases in myocardial oxygen demand occur before the onset of ischemia. Analysis of the pattern of the change in heart rate before episodes of ST segment depression revealed that such increases may occur over a protracted period—up to 30 minutes,[14] although the major increase appears to occur in the immediate 5-minute period before ischemia.[13,14] Indeed, it appears that the likelihood of ischemia is related to the magnitude and duration of increase in heart rate, ranging from a 4% likelihood with a heart rate increase of 5–9 bpm, and lasting less than 10 minutes, to a 60% likelihood when heart rate increases by more than 20 bpm and lasts for more than 40 minutes.[17] It can be estimated that in patients with stable coronary artery disease who have episodes of transient myocardial ischemia during daily living, between 80% and 90% of all ischemic episodes occur with preceding increases in heart rate. When we carefully analyzed heart rate changes related to episodes occurring during the night, a time when physical activity is minimal, we found that the majority of these episodes were also preceded by increases in heart rate, often occurring as a result of awakening from deep sleep, turning in bed, and occasionally due to REM sleep and sleep apnea.[16,18]

Examination of diurnal changes in heart rate, blood pressure, and myocardial contractility, the latter estimated indirectly from changes in plasma epinephrine levels, demonstrates that all of these physiological phenomena, known to be determinants of myocardial oxygen demand, are elevated in the morning hours.[19,20] Indeed, the pattern of their distribution parallels the uneven distribution of ischemia during the day and night (Fig. 1). Thus, it seems reasonable to conclude that an important and perhaps predominant trigger for myocardial ischemia, and for its distribution during the day, is a change in myocardial oxygen demand.

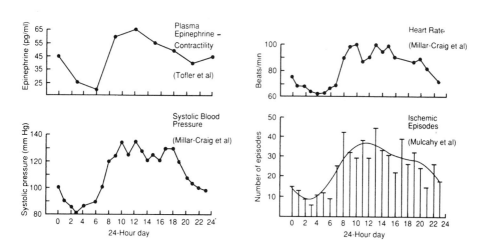

Figure 1: Circadian variation in determinants of myocardial oxygen demand and in transient ischemic activity. (Reproduced with permission of American Heart Journal.[5])

Role of Vasoconstrictor Tone in Myocardial Ischemia

There is much clinical evidence that favors the role of alternative factors as contributors to myocardial ischemia.

Environmental Factors

Psychological stress and exposure to cold are important environmental triggers for precipitation of myocardial ischemia. Hypoperfusion, demonstrated by positron emission tomography, occurred in myocardial regions during mental stress, identical to the defects produced by exercise.[21] Using radionuclide ventriculography during a mentally challenging speech test, it was demonstrated that wall motion abnormalities occurred at a considerably lower rate-pressure product than that achieved during physical exercise, when similar wall motion abnormalities were recorded.[22] Exercise in the cold is known to considerably reduce ischemic threshold,[23] suggesting that there are many occasions during daily living where ischemia can be precipitated at a considerably lower workload than at other times, and implying an important role for changes in coronary vasomotor tone.

These observations are confirmed by studies using ambulatory monitoring which show that ischemia occurs not only during physical activities but also during less demanding activities such as eating, sleeping, or mental stress in the same patients. Analysis of heart rate changes before the onset of ischemic episodes has demonstrated that between 10% and 20% of episodes are not preceded by increases in heart rate[13,14,17,24] There are two possible explanations for these findings. First, although heart rate is taken as a surrogate for myocardial oxygen demand, it is possible that large increases in blood pressure and contractility are responsible for ischemia during episodes where heart rate increases appear to be less than expected. Rozanski et al. have reported that, during mental stress, systolic blood pressure reaches levels similar to those achieved during maximal exercise, whereas heart rate increases prior to ischemia are significantly less than those achieved during exercise.[22] Alternatively, it is likely that changes in the caliber of the coronary arteries, especially at the site of coronary narrowing, may transiently lower coronary blood flow, such that myocardial ischemia can occur with little or no change in myocardial oxygen demand.

Systemic Factors in Precipitating Ischemia

These external triggers for ischemia probably act via several systemic and local mechanisms. Systemic vasoactive mediators appear to

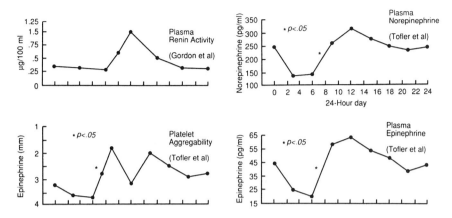

Figure 2: Circadian variation in determinants of myocardial blood flow. (Reproduced with permission of the American Heart Journal.[5])

have a circadian pattern of distribution parallel to changes in myocardial ischemic episodes. For example, catecholamine levels surge in the morning hours, resulting in dramatic increases in circulating levels of norepinephrine and epinephrine.[20] Alpha-adrenergic receptor stimulation with consequent increases in coronary and peripheral tone may thus contribute to the morning increase in ischemic activity. Similarly, another powerful endogenous vasoconstrictor system, the renin-angiotensin system, is activated in the morning hours.[25] Whether this merely constitutes an increase in plasma renin activity and subsequent angiotensin II production, or whether there is a concomitant activation of the tissue renin-angiotensin system is not known. Increases in both the catecholamine levels and plasma renin activity are triggered by assumption of upright posture and exercise.[26] An increase in plasma cortisol in the morning hours acts synergistically with catecholamines in altering vasomotor tone. Thus, vascular tone may be increased in the morning hours, especially after arousal and assumption of upright posture as a result of increases in systemic noradrenergic and renin activities[27] (Fig. 2).

Local Factors and Coronary Vasomotion

Pathological sectioning of atherosclerotic coronary arteries at the sites of severe stenoses exhibits encroachment of the lumen by fibrotic, often calcified, atherosclerotic material in the subintimal space.[28] It was initially suggested that fibrotic coronary arteries were incapable of vasomotion and even the beneficial effects of sublingual nitroglycerin were attributed to its peripheral vasodilatory effects. However, pathophysiological and functional studies have since shown that significant vasoconstriction can occur even in severely stenosed atherosclerotic coronary arteries. Coronary artery sectioning at 5-mm intervals demonstrated that almost three-quarters of all sections had eccentric intimal thickening with preservation of smooth muscle in relatively uninvolved sections of the lumen.[28] Brown and colleagues[29,30] studied epicardial coronary artery diameter changes of atherosclerotic segments at rest and after hand grip exercise (Fig. 3). Despite a 20% increase in heart rate, a 24% increase in blood pressure, and a 66% increase in coronary blood flow, angiographically "normal" appearing segments narrowed by 15–22% and stenosed segments by 5–33%. Other investigators have demonstrated similar paradoxic vasoconstriction in

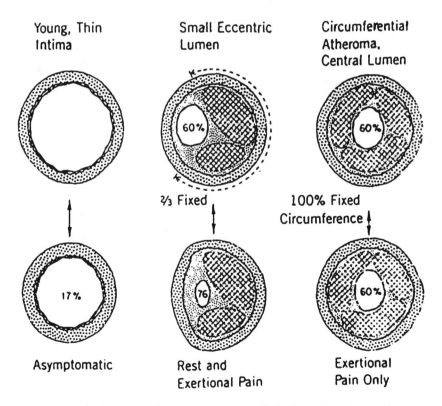

Figure 3: Relationship of lesion structure to clinical presentation in the spectrum of anginal syndromes; effects of ordinary vasoconstriction in normal and diseased coronary arteries. In portions of the vessel with relatively normal wall, transient 10% isovolumetric circumferential shortening may cause dramatic change in lumen caliber. (Adapted with permission of the American Medical Association.[30])

atherosclerotic coronary arteries in vivo during physical exercise, mental stress, and cold-pressor tests.[31–35] Patients with normal coronary arteries, in contrast, vasodilate epicardial coronary arteries during such physiological stimuli.

It appears, therefore, that both vasoconstriction of stenotic coronary arteries, leading to dramatic increases in coronary vascular resistance, and increases in myocardial oxygen demand occur simultaneously during physical exercise or during emotional and environmental stresses. It is also clear from the ambulatory monitoring data, which show considerable variation in heart rate threshold during different ischemic episodes, that the variation in vasomotor tone does indeed influence the ease with which myocardial ischemic episodes occur at different times in the same individual.

Circadian Variation in Vasoconstrictor Tone

To investigate the possibility that the morning surge in myocardial ischemia is due not only to an increase in demand at that time but also to an increase in coronary vasoconstrictor tone, we exercised patients with myocardial ischemia and coronary artery disease four times during the day and measured their ischemic threshold (heart rate at onset of ST segment depression).[36] The ischemic threshold, or the ease with which myocardial ischemia developed, was lower in the morning and at night compared to other times of the day. This was paralleled by a simultaneous circadian variation in postischemic forearm vascular resistance, which was increased in the morning and at night compared to other times of the day (Fig. 4). These findings indicate that there is a circadian variation in coronary vascular resistance such that it is lower in the early morning (8 AM) and at night (9 PM) compared to other times of the day and that this change in vascular resistance is likely to be generalized and affect other vascular beds.

Other studies also support the concept of an elevated coronary vascular resistance in the early morning and at night compared to other times of the day. Examination of heart rate at onset of spontaneously occurring episodes of ischemia during ambulatory monitoring has demonstrated that the heart rate at onset of ischemic episodes at night is significantly lower than the heart rate at onset of ischemia during the daytime.[37] Cardiac pacing during ischemia in patients with rest angina demonstrated that the heart rate at the onset of ischemia was lower at night than it was during the day.[38] Angiography performed in patients with normal coronary arteries in the early morning (7 AM) and later in the day demonstrated narrower coronary arteries in the morning.[39] Under basal conditions, coronary blood flow was significantly

Circadian Variation in Post-Ischemic Forearm Vascular Resistance and Flow and Ischemic Threshold

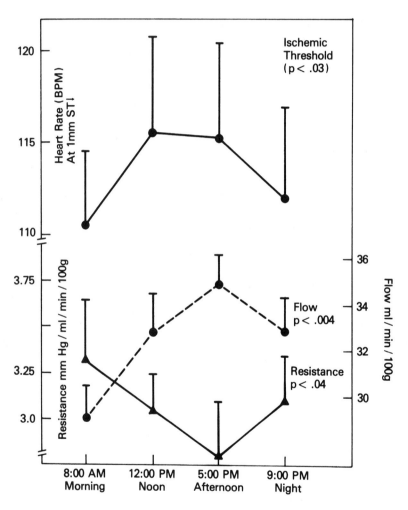

Figure 4: Figure showing circadian variation in ischemic threshold, post-ischemic forearm vascular resistance, and blood flow in patients with stable coronary artery disease. Ischemic threshold is lower in the morning waking hours and at night compared with during the day and corresponds with higher forearm vascular resistance in the early morning and at night. (Reproduced with permission of American Heart Association.[59])

greater in the afternoon than in the morning in dogs.[40] Thus, these separate pieces of evidence support the concept that the coronary vasculature is responsive to vasomotor influences and that the constrictor tone predominates at night and in the early morning hours compared to the rest of the day.

Although much data exist on changes in epicardial coronary vascular resistance with physiological stress, there is increasing evidence to suggest that microvascular coronary tone is also abnormal during stress in patients with atherosclerosis. Thus, patients with coronary atherosclerosis without a critical narrowing of the epicardial coronary artery or those exposed to risk factors for atherosclerosis[41,42] appear to have depressed vasodilation in response to mental stress, cold-pressor test, or cardiac pacing, suggesting impairment of microvascular vasodilation during stress.[33,43,44]

Sympathetic Nervous System Activation and Vasomotor Tone

The role of systemic vasoconstrictor influences, in particular sympathetic nervous system activation, has been investigated in recent years. In isolated human epicardial coronary arteries, both α-1 and α-2 adrenoceptors are involved in norepinephrine-induced constriction.[45] Although the role of α-1 and α-2 adrenoceptors in the human coronary microcirculation is not fully clarified at the present time, the exercise-, cold-pressor-, and mental stress-induced constriction of atherosclerotic coronary arteries as described above can be reversed by α-blockade.[46] In another study, exercise-induced ST segment depression was delayed and reduced by intracoronary phentolamine.[47] In atherosclerotic patients, mental stress-induced increases in coronary vascular resistance, suggesting microvascular constriction, were also reversed by intracoronary phentolamine.[48]

To investigate whether the circadian pattern of change in coronary and systemic vascular resistance described earlier is also determined by changes in circulating norepinephrine levels and thus alpha-adrenergic sympathetic tone, we performed a study to investigate the effect of phentolamine on the circadian variation of forearm vascular resistance, previously shown to be higher in the early morning hours compared to later times of the day.[49] Phentolamine produced a more dramatic reduction in forearm vascular resistance during the morning hours compared to other times of the day, leading to abolition of the circadian variation in vascular resistance. A nonspecific vasodilator, sodium nitroprusside, also reduced forearm vascular resistance, but its effect was similar at all times of the day.

The therapeutic implications of these findings are still under investigation. There appears to be no convincing evidence that α-adrenergic blockade is of therapeutic benefit in patients with variant angina, where alterations in coronary vasomotor tone are severe.[50–55] Increase in exercise capacity has nevertheless been reported in patients with stable coronary artery disease with an α-1 adrenergic antagonist, indoramin.[56] Whether α-1 blockade will be helpful in attenuating the morning increase in vascular resistance and thus the surge in ischemic activity needs to be investigated.

Pathophysiology of Epicardial Coronary Artery Vasomotion

There may be several reasons for exercise-induced vasoconstriction or "collapse" of atherosclerotic coronary arteries. According to hydrodynamic principles described by Bernouille, there is a fall in pressure across a significantly narrowed lesion due to turbulence. This fall in pressure across a compliant, eccentrically narrowed segment may lead to partial collapse of the lesion. Nevertheless, it now appears that other, locally active factors contribute substantially to this phenomenon. The importance of increased α-adrenergic tone has been discussed earlier. Over the past decade, the role of the endothelium in explaining these phenomena has been clarified and is discussed below.

Physiology of the Endothelium

The endothelium plays a vital role in all aspects of human biology, regulating the actions of the cardiovascular, gastrointestinal, respiratory, genitourinary and neurological systems, and its role in thrombosis is also crucial.[57] In the cardiovascular system, its functions include, among others, transport and permeability of solutes and fluids, regulation of lipids, participation in immune and inflammatory responses, control of cell growth and proliferation, and a role in angiogenesis. Its role both in the control of vascular smooth muscle tone and in platelet function will be discussed in detail here.

The endothelium regulates tone and platelet aggregation by releasing an array of mediators (Fig. 5). The dilators include endothelium-derived relaxing factors (nitric oxide and endothelium-derived hyperpolarizing factor) and prostacyclin (PGI_2).[58,59] Constricting factors released by the endothelium are arachidonic acid metabolites such as thromboxane A_2, and endoperoxidase, or a potent 21-amino acid peptide called endothelin.[60] The endothelium is also a major site for conversion of angiotensin I to angiotensin II by the membrane-bound an-

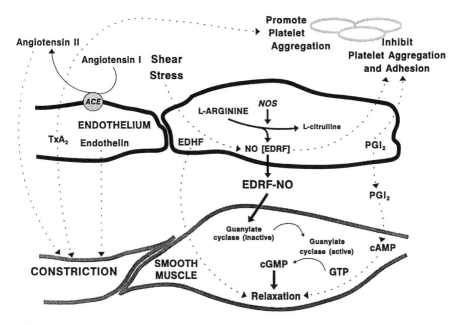

Figure 5: Schematic diagram showing mediators released from the endothelial cells that impact on coronary/vascular tone and platelet aggregability.

giotensin-converting enzyme (ACE). In addition, it is now believed that the vascular wall is able to express renin and synthesize angiotensin II from angiotensin I, thus making angiotensin II another endothelium-dependent contracting factor.[61]

The role of contracting factors in the dynamic control of vasomotor tone and their influence during physiological stresses is poorly understood. Although all of these agents are powerful vasoconstrictors, are tonically released by the endothelium, and can be further stimulated by a variety of pharmacological stimuli and disease states, their action appears to be counterbalanced by endothelium-derived relaxing factors (EDRFs).

Endothelium-Derived Nitric Oxide

The endothelium releases relaxing factors (EDRFs), of which the most important appears to be nitric oxide (NO), or a compound closely akin to nitric oxide. NO is tonically produced in picomole quantities by the endothelium from L-arginine, by the action of a constitutive calcium and calmodulin-dependent enzyme, nitric oxide synthase.[58,59,62] In blood, NO has a high affinity for hemoglobin and can be rapidly inactivated by free hemoglobin. NO is also rapidly oxidized to higher oxides of nitrogen by oxygen free radicals. It can also nitrosate sulfhydryl group-containing molecules such as glutathione, cysteine, and albumin and produce carriers or sinks for NO in biological fluids.[63]

A variety of pharmacological and physiological factors can promote release of NO.[64] These include acetylcholine, substance P, bradykinin, ADP, ATP, thrombin, serotonin, and histamine. Physiological factors that stimulate NO production by as yet undetermined mechanisms include increasing shear stress over the lumen surface and by aggregating platelets.[59,65,66]

It is now known that NO is released tonically from the vascular endothelium. Thus, inhibition of NO synthesis by L-N^G monomethyl arginine (L-NMMA) causes constriction of conductance vessels and reduction in blood flow confirming basal formation of NO from both the epicardial and microvessels.[67] Basal activity of NO is much greater from arteries and arterioles when compared with veins. Stimulated release of NO is also substantially greater from arteries than veins, as demonstrated by the effects of acetylcholine on human internal mammary arteries compared to saphenous veins.[68]

Implications for Vascular Tone

Understanding of the balance between shear stress-induced endothelium-dependent relaxation and pressure or stretch-induced contraction (myogenic contraction) may help explain some of the observations made in coronary arteries during physiological stress. Normal coronary and indeed peripheral arterial tone is elevated in the morning hours. This might not only be due to increases in circulating vasoconstrictor hormones such as norepinephrine and plasma renin activity, but due to increases in pressure-induced endothelium-dependent and direct smooth muscle contractions. The latter may be stimulated by the sharp increase in blood pressure that occurs during arousal, awakening, and assumption of the upright posture. This constriction likely increases shear stress and promotes production of EDRF and prostacyclin which, in turn, counteracts the vasoconstriction. In the presence of endothelial dysfunction (see below), where EDRF activity is reduced, this vasoconstriction may remain unopposed because of the deficient shear stress-induced release of EDRF. Thus, under such circumstances, shear stress will increase more dramatically in diseased coronary arteries and may, in fact, contribute to plaque rupture that appears to be more prominent in the morning hours.

Endothelial Dysfunction

Because the endothelium participates in so many biological processes, the term endothelial dysfunction refers to any imbalance between relaxing and contracting factors, between pro-and anticoagulant mediators, or between growth inhibiting and growth promoting factors. This imbalance will, for example, result in depressed activity of NO and prostacyclin which are not only smooth muscle-relaxing factors, but are also anticoagulants and antigrowth in their action. Endothelial dysfunction may not actually be a result of pathologically visible physical damage to the cells but may represent purely a phenotypic modulation of the endothelial cell to activation or injury.

Atherosclerosis and Hypercholesterolemia

The endothelium appears to be a chief target for cardiovascular risk factors that accompany atherosclerosis. Cholesterol-fed animal models have demonstrated depressed endothelial production of NO at rest and on stimulation.[69–71] These observations have been confirmed in humans with hypercholesterolemia both with and without established atherosclerosis.[72–76] Acetylcholine in low doses causes endothelium-dependent epicardial vessel vasodilation in normal smooth coronary arteries which is abolished or converted to constriction in patients with hypercholesterolemia or atherosclerosis. Interestingly, the abnormalities of endothelium-dependent vasodilation are observed in both epicardial and conductance vessels that are directly involved in atherosclerosis and also in microvessels that do not develop morphological changes in atherosclerosis.[77] It is not clear how hypercholesterolemia causes endothelial dysfunction but it appears that oxidized LDL is a key determinant. Generation of endothelial superoxides in hypercholesterolemic vessels has been demonstrated and rapid inactivation of NO by superoxide anions may be one cause of depressed endothelial function observed in hypercholesterolemia.[78–80] Forearm vasodilation response to acetylcholine, an endothelium-dependent vasodilator, is depressed in patients with hypercholesterolemia, but endothelium-independent responses to sodium nitroprusside are preserved, demonstrating reduced stimulated release of EDRF in hypercholesterolemia.[74–76] Similar observations have been made in coronary epicardial arteries and in microvessels.

Endothelial Function in Hypertension, Diabetes, Aging, and Heart Failure

Forearm vasodilation with acetylcholine is depressed in hypertension whereas endothelium-independent responses are preserved.[81,82] L-N^G monomethyl L arginine (L-NMMA), the inhibitor of NO production, had an insignificant effect on resting blood flow in these patients, suggesting that both basal and stimulated release of NO is depressed in hypertensive patients. Similar results have been reported in the coronary vasculature of patients with hypertension and left ventricular hypertrophy.[83] Diabetes also results in similar endothelial dysfunction. In heart failure, both endothelium-dependent and endothelium-independent vascular responses appear to be depressed.[84] Depressed endothelium-dependent vasodilation with acetylcholine was partially restored by inhibition of cyclooxygenase, suggesting an increased production of cyclooxygenase enzyme-dependent constricting factor in heart failure. Abnormal endothelium-dependent epicardial reactivity has also been reported in older patients compared to the young with normal coronary arteries.[85,86]

Exposure to multiple risk factors for coronary artery disease such as hypercholesterolemia, hypertension, diabetes, age, and smoking can compound the injury to the endothelium.[67,73,87] We have demonstrated constriction of epicardial coronary arteries to intracoronary acetylcholine and depressed flow response, indicating microvascular and epicardial vessel endothelial dysfunction in patients with multiple coronary risk factors.[67,88] The magnitude of endothelial dysfunction of the coronary and peripheral vasculature appears to correlate with the number of risk factors. Not only is stimulated release of NO depressed, but basal release appears to be diminished in patients exposed to risk factors despite angiographically normal coronary arteries.

Consequences of Endothelial Dysfunction

Because the endothelium releases EDRF in response to physiological stimuli such as shear stress, we investigated whether endothelial dysfunction resulted in reduced shear-induced dilation of the coronary vasculature. We were able to demonstrate that patients with reduced vasodilation in response to acetylcholine also had depressed dilation in response to atrial pacing.[44] In a subsequent study, we also demonstrated depressed vasodilation in response to acetylcholine and cardiac pacing in patients with risk factors for atherosclerosis[89]; these patients also had a depressed response to L-NMMA, suggesting that there is impaired pharmacological and physiological activity of NO in these patients. Patients with risk factors and endothelial dysfunction failed to dilate the epicardial coronary arteries in response to pacing whereas those without risk factors dilated epicardial coronary arteries in response to pacing as well as with acetylcholine. Similar responses have been reported with several other physiological stimuli such as exercise, mental stress, cold, and increased flow.[33,43,90] These stimuli cause epicardial vessel vasodilation in patients with angiographically smooth, normal coronary arteries. However, in patients with atherosclerosis and in segments with angiographic irregularities, these same stimuli lead to vasoconstriction. These studies further demonstrate that segments which constrict in response to physiological stimuli also constrict in response to acetylcholine, whereas normal smooth segments of coronary arteries that dilate during physiological stimulation also dilate in response to acetylcholine. Thus, there is a strong correlation between endothelium-dependent dilation and physiological vasomotion in human coronary vessels.

Platelets and Coronary Arterial Vasomotion

Cyclical reductions in coronary blood flow were demonstrated to occur when a fixed critical stenosis was created in a dog coronary artery.[91,92] These changes were reversed by aspirin and other platelet inhibitors and were shown to be due to deposition of platelet aggregates at the stenotic site, resulting either in temporary reduction in flow due to intermittent plugging of the vessel lumen or release of locally active platelet-derived vasoconstrictors such as thromboxane A_2 and serotonin.

Clinical investigations have demonstrated that platelet activation occurs in the morning hours and appears to be triggered by awakening and assumption of upright posture[93] Platelet activation can also be demonstrated when coronary blood flow is increased by atrial pacing across a stenotic coronary artery, but not in those with nonstenosed coronary arteries.[94] Physiological events such as physical and mental stress can also produce activation of platelets. The mechanisms underlying shear stress-induced increases in platelet activation or platelet activation that occurs with awakening in the morning are unknown. It is possible that the catecholamine level increase in the mornings increases platelet activation by stimulating surface α-2 adrenoceptors that are known to increase platelet aggregation by reducing intracellular levels of cyclic AMP.

Endothelium-Platelet Interactions

As described earlier, both endothelium-derived NO and prostacyclin contribute importantly to platelet aggregation and adhesion.[95,96] NO inhibits both aggregation and adhesion by increasing platelet levels of cyclic GMP, whereas prostacyclin primarily inhibits platelet aggregation by increasing platelet cyclic AMP levels. Tonic release of these mediators from normal endothelium is important in preventing platelet adhesion and subsequent thrombus formation in normal blood vessels. Deficient release of endothelium-derived antiaggregatory and antiadhesive agents in atherosclerosis or in those with risk factors for atherosclerosis can potentially increase the chances of platelet aggregation over atherosclerotic plaques.

To investigate whether these effects of endothelium-derived relaxing factors on platelets are important in vivo, we tested the effect of intracoronary acetylcholine on platelet cyclic GMP levels in blood collected from the coronary sinus. There was a notable increase in platelet cyclic GMP levels, but this was significant only in patients who had no evidence of endothelial dysfunction, confirming that endothelium-derived NO released by acetylcholine can lead to inhibition of platelet aggregation by increasing platelet cyclic GMP levels, this effect being less in patients with endothelial dysfunction.[97]

It is not known whether platelets aggregate on atherosclerotic plaques in vivo; however, it can be hypothesized that endothelial dysfunction at sites of atherosclerosis could promote platelet aggregation and plugging, causing local release of vasoconstrictors, and thus accentuating the magnitude of constriction.

Long-Term Consequences of Endothelial Dysfunction

Reduced activity of NO in the vascular wall can promote the atherogenic process.[98–100] Animals fed a high-cholesterol diet and L-NMMA (which inhibits endothelial production of NO) had significantly more rapid development of intimal thickening than those fed a high-cholesterol diet alone, suggesting that endothelium-derived NO prevents the development and progression of atherosclerosis. The exact mechanisms underlying this effect of NO are unknown, but could be multifactorial. NO inhibits cell growth and proliferation.[98] It is also able to scavenge superoxide anions that are constantly generated in the endothelium. Decreased NO activity would result in excess superoxide activity and in the presence of hypercholesterolemia would lead to oxidation of LDL and accumulation of oxidized LDL in the vessel wall, which acts as a trigger, together with macrophages in development of the atherosclerotic plaque. Less NO will limit the antiadhesive properties of the endothelial cell layer and lead to formation of platelet aggregates and release of platelet-derived growth factors (PDGF) which can also, by their mitogenic effects, promote atherosclerosis. Lower NO activity may enhance the activity of other pro-growth endothelium-derived agents such as endothelin and angiotensin II. Thus, it appears that depression of endothelial nitric oxide function can not only acutely lead to abnormalities of vasomotion of vessels in response to stress and increase the likelihood of thrombosis and platelet aggregation, but also accelerate or promote development of atherosclerosis.

Therapeutic Implications of Endothelial Dysfunction

There are several potential therapeutic avenues for treating endothelial dysfunction:

1. *Substitution of mediators released by endothelium:* These include t-PA, stable analogs of prostacyclin, and nitric oxide donors[101,102] such as nitroglycerin or nitroprusside that yield NO directly in smooth muscles or in the platelets.

2. *Precursors of NO production:* There is mounting evidence that administration of L-arginine can reverse the abnormal endothelium-dependent responses in hypercholesterolemia,[75,103] although this effect has not been demonstrated in some studies of patients with hypertension and hypercholesterolemia.[104,105] In cholesterol-fed animals, long-term administration of L-arginine is believed to reduce the rate of development of the atherosclerotic plaque.[106]

3. *Inhibitors of endothelium-dependent contracting factors:* These include ACE inhibitors that not only antagonize the constriction due to angiotensin II, but also increase endothelium-dependent relaxant effects of the endogenously present bradykinin, which is normally inactivated by endothelial angiotensin-converting enzyme.[107] Other drugs in this category include specific angiotensin II receptors antagonists, thromboxane A_2 synthesis inhibitors and receptor antagonists, and endothelin antagonists.

4. *Free radical scavengers:* Increased oxidative stress due to vascular superoxide anion activity may be responsible for the reduced activity of NO in certain pathological conditions.[79,80] Superoxide dismutase,[108] oxypurinol, vitamin E, and other antioxidants play a role in these conditions in reversing the endothelial abnormality. These agents are currently under investigation.

5. *Lipid-lowering drugs:* Although now widely accepted as being able to reduce the rate of progression of atherosclerosis, the disproportionately dramatic effect of these agents in reducing coronary events remains unexplained.[109] Animal studies have demonstrated restoration of endothelial function in hypercholesterolemic rabbits given lipid-lowering drugs.[110] These findings have recently been confirmed in the human coronary vasculature, where endothelium-dependent reactivity of acetylcholine was normalized by pravastatin treatment.[111]

Thus, understanding the physiology and pathophysiological role of the vascular endothelium as a modulator of smooth muscle vasomotion, intravascular thrombosis, and intimal growth will greatly enhance future possibilities for treatment of the atherosclerotic process and its effects on vascular function.

The Role of Vasomotor Tone in Acute Coronary Syndromes

In contrast to its role in stable angina pectoris, changes in vasomotor tone appear to play a secondary role in the pathogenesis of acute coronary syndromes. Unstable angina pectoris, myocardial infarction, and many cases of sudden cardiac death are believed to be due to plaque rupture or fissuring and formation of intraluminal thrombus with acute accentuation of lesion severity.[12] Studies using nitroglycerin, a powerful dilator of smooth muscle, have demonstrated beneficial effects in unstable angina, suggesting that superimposed vasoconstriction after disruption of the plaque is an important component of blood flow reduction to the ischemic territory.[112] Similar evidence has been gathered in the setting of acute MI. Hackett and colleagues have demonstrated that the culprit lesion was more responsive to intracoronary isosorbide dinitrate than other areas in the vascular tree in patients after myocardial infarction.[113] Thus, it appears that superimposed vasoconstriction at the site of plaque destruction occurs during these acute cardiac syndromes.

While there is little evidence to support a primary causative association between alterations in coronary vasomotor tone and plaque disruption, coronary spasm can lead to acute MI[114,115] and cardiac death,[116] and it is possible that coronary vasoconstriction, by increasing shear stresses, can indirectly lead to plaque rupture with its attendant complications. Furthermore, in the unusual circumstance of extreme alterations in coronary vasomotor tone (coronary spasm), which can occur at the sites of minimal stenoses as well as more severe lesions, prolonged occlusion can result in myocardial necrosis and life-threatening arrhythmias in the absence of plaque rupture. This "syndrome" is an example of the ability of even the most severe "fixed" stenosis to vasoconstrict, and the underlying pathophysiological mechanism is readily confirmed by the rapid response to nitrates and calcium antagonist agents. Furthermore, it has been shown that spontaneous occlusive episodes are more likely to occur in the early morning hours than at other times of the day,[117] in sharp contrast with transient ischemic episodes in stable angina, which occur predominantly during activity hours, and are rare in the early morning hours. This observation of a predilection to spontaneous episodes of "spasm" in the early morning hours relates to what we know of the activity of coronary vasomotion, which is increased at this time compared to the activity hours in normal circumstances; however, it is fortunately rarely expressed in such an extreme way.

Conclusion

Over the past decade, dramatic advances have been made in our understanding of the behavior of the coronary artery, at both the vascular and the cellular level. Issues of simple "demand versus supply" are being largely superseded with the development of an increasingly clearer understanding of the astonishing complexity of dynamic activity at the level of the coronary artery, both in health and in disease. Multiple interactive components at the endothelial, vascular, and hemostatic level combine to ensure the normal functioning of the coronary tree, but may also cause and propagate the atherosclerotic process in the right environment and can contribute, and even lead, to the endpoints of coronary disease. These endpoints are more likely to occur in the morning waking hours than at other times of the day and at a time when coronary artery tone is known to be increased. Alterations in coronary vasomotor tone play an important part in these acute pathophysiological processes in propagating any disturbance, if not as a direct causative mechanism.

References

1. Mulcahy D, Keegan J, Cunningham D, et al: Circadian variation of total ischaemic burden and its alteration with antianginal agents. Lancet 1988; ii:755–759.
2. Rocco MB, Barry J, Campbell S, et al: Circadian variation of transient myocardial ischemia in patients with coronary artery disease. Circulation 1987; 75:395–400.
3. Parker JD, Testa MA, Jimenez AH, et al: Morning increase in ambulatory ischemia in patients with stable coronary artery disease. Importance of physical activity and increased cardiac demand. Circulation 1994; 89:604–614.
4. Quyyumi AA, Mockus L, Wright C, Fox KM: Morphology of ambulatory ST segment changes in patients with varying severity of coronary artery disease: investigation of the frequency of nocturnal ischaemia and coronary spasm. Br Heart J 1985; 53:186–193.
5. Quyyumi AA: Circadian rhythms in cardiovascular disease. Am Heart J 1990; 120:726–733.
6. Raeder EA, Hohnloser SH, Graboys TB, Podrid P, Lampert S, Lown B: Spontaneous variability and circadian distribution of ectopic activity in patients with malignant ventricular arrhythmia. J Am Coll Cardiol 1988; 12:656–661.

7. Muller JE, Stone PH, Turi ZG, et al: Circadian variation in the frequency of onset of acute myocardial infarction. N Engl J Med 1985; 313:1315–1322.

8. Willich SN, Linderer T, Wegscheider K, Leizorowicz A, Alamercery I, Schroder R: Increased morning incidence of myocardial infarction in the ISAM study: absence with prior beta-adrenergic blockade. Circulation 1989; 80:853–858.

9. Lampert R, Rosenfeld L, Batsford W, Lee F, McPherson C: Circadian variation of sustained ventricular tachycardia in patients with coronary artery disease and implantable cardioverter defibrillators. Circulation 1994; 90:241–247.

10. Muller JE, Ludmer PL, Willich SN, et al: Circadian variation in the frequency of sudden cardiac death. Circulation 1987; 75:131–138.

11. DeWood MA, Spores J, Notske R, Mouser LT, Borroughs R, Golden MS, Lang HT: Prevalence of total coronary occlusion during the early hours of transmural myocardial infarction. N Engl J Med 1980; 303:897–902.

12. Davies MJ, Thomas AC: Thrombosis and acute coronary-artery lesions in sudden cardiac ischemic death. N Engl J Med 1984; 310:1137–1140.

13. Panza JA, Diodati JG, Callahan TS, Epstein SE, Quyyumi AA: Role of increases in heart rate in determining the occurrence and frequency of myocardial ischemia during daily life in patients with stable coronary artery disease. J Am Coll Cardiol 1992; 20:1092–1098.

14. McLenachan JM, Weidinger FF, Barry J, et al: Relations between heart rate, ischemia, and drug therapy during daily life in patients with coronary artery disease. Circulation 1991; 83:1263–1270.

15. Deedwania PC, Nelson JR: Pathophysiology of silent myocardial ischemia during daily life: hemodynamic evaluation by simultaneous electrocardiographic and blood pressure monitoring. Circulation 1990; 82:1296–1304.

16. Quyyumi AA, Wright C, Mockus LJ, Fox KM: Mechanisms of nocturnal angina pectoris: importance of increased myocardial oxygen demand in patients with severe coronary artery disease. Lancet 1984; 2:1207–1209.

17. Andrews TC, Fenton T, Toyosaki N, et al: Subsets of ambulatory myocardial ischemia based on heart rate activity: circadian distribution and response to anti-ischemic medications. Circulation 1993; 88:92–100.

18. Quyyumi AA, Efthimiou J, Quyyumi A, Mockus LM, Spiro SG, Fox KM: Nocturnal angina: precipitating factors in patients with coronary artery disease and those with variant angina. Br Heart J 1986; 56:346–352.

19. Millar-Craig MW, Bishop CN, Raftery EB: Circadian variation of blood pressure. Lancet 1978; i:795–797.

20. Turton MB, Deegan T: Circadian variations of plasma catecholamine, cortisol, and immunoreactive insulin concentrations in supine subjects. Clin Chim Acta 1974; 55:389–397.

21. Deanfield JE, Selwyn AP, Chierchia S, et al: Myocardial ischaemia during daily life in patients with stable angina: its relation to symptoms and heart rate changes. Lancet 1983; i:753–758.

22. Rozanski A, Bairey N, Krantz DS, et al: Mental stress and induction of silent myocardial ischemia in patients with coronary artery disease. N Engl J Med 1988; 318:1005–1012.

23. Epstein SE, Stampfer M, Beiser GD, et al: Effects of a reduction in environmental temperature on the circulatory response to exercise in man: implications concerning angina pectoris. N Engl J Med 1969; 280:7–11.

24. Mulcahy D, Keegan J, Fox KM: Characteristics of silent and painful ischaemia during ambulatory monitoring in patients with coronary arterial disease. Int J Cardiol 1990; 28:377–379.

25. Gordon RD, Wolfe LK, Island DP, Liddle GW: A diurnal rhythm in plasma renin activity in man. J Clin Invest 1966; 45:1587–1592.

26. Brezinski DA, Tofler GH, Muller JE, et al: Morning increase in platelet aggregability: association with assumption of the upright posture. Circulation 1988; 78:35–40.

27. Weitzman ED, Fukushima D, Nogeire C, Roffwarg H, Gallagher TF, Hellman L: Twenty-four hour pattern of the episodic secretion of cortisol in normal subjects. J Clin Endocrinol Metab 1971; 33:14–22.

28. Quyyumi AA, Al-Rufaie H, Olsen EGJ, Fox KM: Coronary anatomy in patients with varying manifestations of three vessel coronary artery disease. Br Heart J 1985; 54:362–366.

29. Brown BG: Coronary vasospasm: observations linking the clinical spectrum of ischemic heart disease to the dynamic pathology of coronary atherosclerosis. Arch. Intern Med 1981; 141:716–722.

30. Brown BG, Lee AB, Bolson EL, Dodge HT: Reflex constriction of significant coronary stenosis as a mechanism contributing to ischemic left ventricular dysfunction during isometric exercise. Circulation 1984; 70:18–24.

31. Nabel EG, Selwyn AP, Ganz P: Paradoxical narrowing of atherosclerotic coronary arteries induced by increases in heart rate. Circulation 1990; 81:850–859.

32. Nabel EG, Ganz P, Gordon JB, Alexander RW, Selwyn AP: Dilation of normal and constriction of atherosclerotic coronary arteries caused by the cold pressor test. Circulation 1988; 77:43–52.

33. Gordon JB, Ganz P, Nabel EG, et al: Atherosclerosis influences the vasomotor response of epicardial coronary arteries to exercise. J Clin Invest 1989; 83:1946–1952.

34. Yeung AC, Vekshtein VI, Krantz DS, et al: The effect of atherosclerosis on the vasomotor response of coronary arteries to mental stress. N Engl J Med 1991; 325:1551–1556.

35. Gage JE, Hess OM, Murakami T, Ritter M, Gramm J, Krayenbuehl HP: Vasoconstriction of stenotic coronary arteries during dynamic exercise in patients with classic angina pectoris: reversibility by nitroglycerin. Circulation 1986; 73:865–876.

36. Quyyumi AA, Panza JA, Diodati JG, Lakatos E, Epstein SE: Circadian variation in ischemic threshold. A mechanism underlying the circadian variation in ischemic events. Circulation 1992; 86:22–28.

37. Benhorin J, Banai S, Moriel M, et al: Circadian variation in ischemic threshold and their relation to the occurrence of ischemic episodes. Circulation 1993; 87:808–814.

38. Figueras J, Cinca J, Balda F, Moya A, Rius J: Resting angina with fixed coronary artery stenosis: nocturnal decline in ischemic threshold. Circulation 1986; 74:1248–1254.

39. Yasue H, Omote S, Takizawaw A, Nagao M, Miwa K, Tanaka S: Circadian variation of exercise capacity in patients with Prinzmetal's variant angina: role of exercise-induced coronary arterial spasm. Circulation 1979; 59: 938–948.

40. Fugita M, Franklin D: Diurnal changes in coronary blood flow in conscious dogs. Circulation 1987; 76:488–491.
41. Seiler C, Hess OM, Buechi M, Suter TM, Krayenbuehl HP: Influence of serum cholesterol and other coronary risk factors on vasomotion of angiographically normal coronary arteries. Circulation 1993; 88:2139–2148.
42. Zeiher AM, Schächinger V, Hohnloser SH, Saurbier B, Just H: Coronary atherosclerotic wall thickening and vascular reactivity in humans. Circulation 1994; 89:2525–2532.
43. Zeiher AM, Drexler H, Wollschläger H, Just H: Endothelial dysfunction of the coronary microvasculature is associated with impaired coronary blood flow regulation in patients with early atherosclerosis. Circulation 1991; 84:1984–1992.
44. Quyyumi AA, Cannon RO, Panza JA, Diodati JG, Epstein SE: Endothelial dysfunction in patients with chest pain and normal coronary arteries. Circulation 1992; 86:1864–1871.
45. Heusch G: α-adrenergic mechanisms in myocardial ischemia. Circulation 1990; 81:1–13.
46. Mudge GH, Grossman W, Mills RM Jr, Lesch M, Braunwald E: Reflex increase in coronary vascular resistance in patients with ischemic heart disease. N Engl J Med 1976; 295:1333–1337.
47. Berkenboom GM, Abramowicz M, Vandermoten P, Degre SG: Role of alpha-adrenergic coronary tone in exercise-induced angina pectoris. Am J Cardiol 1986; 57:195–198.
48. Dakak N, Quyyumi AA, Eisenhofer G, Goldstein DS, Cannon RO: Effects of cardiac α-adrenergic blockade on coronary vasomotor responses and cardiac norepinephrine kinetics during mental stress in patients with coronary artery disease. Circulation 1993; 88:I-494.
49. Panza JA, Epstein SE, Quyyumi AA:Circadian variation in vascular tone and its relation to alpha sympathetic vasoconstrictor activity. N Engl J Med 1991; 325:986–990.
50. Yasue H, Touyama M, Kato H, Tanaka S, Akiyama F: Prinzmetal's variant form of angina as a manifestation of alpha-adrenergic receptor-mediated coronary artery spasm: documentation of coronary arteriography. Am Heart J 1976; 91:148–155.
51. Raizner AE, Chahine RA, Ishimori T, Verani MS, Zacca N, Jamal N, Miller RR, Luchi RJ: Provocation of coronary artery spasm by the cold pressor test. Circulation 1980; 62:925–932.
52. Tzivoni D, Keren A, Benhorin J, Gottlieb S, Atlas D, Stern S: Prazosin therapy for refractory variant angina. Am Heart J 1983; 105:252–266.
53. Robertson RM, Bernard YD, Carr RK, Robertson D: Alpha-adrenergic blockade in vasotonic angina: lack of efficacy of specific alpha 1-receptor blockade with prazosin. J Am Coll Cardiol 1983; 2:1146–1150.
54. Winniford MD, Filipchuk N, Hillis LD: Alpha-adrenergic blockade for variant angina: a long-term, double-blind randomized trial. Circulation 1983; 67:1185–1188.
55. Chierchia S, Davies G, Berkenboom G, Crea F, Crean P, Mersi A: α-adrenergic receptors and coronary spasm: an elusive link. Circulation 1984; 69:8–14.
56. Collins P, Sheridan D: Improvement in angina pectoris with alpha adrenoceptor blockade. Br Heart J 1985; 53:488–492.

57. Moncada S, Higgs A: The L-arginine-nitric oxide pathway. N Engl J Med 1993; 329:2002–2012.
58. Furchgott RF, Zawadski JV: The obligatory role of endothelial cells in the relaxation of arterial smooth muscle by acetylcholine. Nature 1980; 288: 373–376.
59. Vanhoutte PM, Shimokawa H: Endothelium-derived relaxing factor and coronary vasospasm. Circulation 1989; 80:1–9.
60. Yanagisawa M, Kurhara H, Kimura S, et al: A novel potent vasoconstrictor peptide produced by vascular endothelial cells. Nature 1988; 332:411–415.
61. Dzau VJ: Vascular wall renin angiotensin pathway in control of the circulation. Am J Med 1984; 77:31–36.
62. Snyder SH, Bredt DS: Biological roles of nitric oxide. Sci Am 1992; 266: 68–71, 74–77.
63. Stamler JS, Simon DI, Osborne JA, et al: S-nitrosylation of proteins with nitric oxide: synthesis and characterization of biologically active compounds. Proc Natl Acad Sci USA 1992; 89:444–448.
64. Vanhoutte PM: Endothelium and control of vascular function. Hypertension 1989; 13:658–667.
65. Holtz J, Forstermann U, Pohl U, Giesler M, Bassenge E: Flow-dependent, endothelium-mediated dilatation of epicardial coronary arteries in conscious dogs: effects of cyclooxygenase inhibition. J Cardiovasc Pharmacol 1984; 6:1161–1169.
66. Rubanyi GM, Romero JC, Vanhoutte PM: Flow-induced release of endothelium-derived relaxing factor. Am J Physiol 1986; 250:H1145–1149.
67. Quyyumi AA, Dakak N, Arora S, Gilligan GM, Johnson GB, Panza JA, Andrews NP, Cannon RO: Effect of inhibition of nitric oxide synthesis in the human coronary circulation. J Am Coll Cardiol 1994; 1A:427A.
68. Lüscher TF, Diederich D, Siebenmann R, et al: Difference between endothelium-dependent relaxation in arterial and in venous coronary bypass grafts. N Engl J Med 1988; 319:462–467.
69. Takahasi M, Yui Y, Yasumoto H, Aoyama T, Morishita H, Hatori R, Kiawai C: Lipoproteins are inhibitors of endothelium-dependent relaxation of rabbit aorta. Am J Physiol 1990;258(Heart Circ Physiol 27):H1–H8.
70. Bossaller C, Yamamoto H, Lichtlen PR, Henry PD: Impaired cholinergic vasodilation in the cholesterol-fed rabbit in vivo. Basic Res Cardiol 1987; 82:396–404.
71. Osborne JA, Siegman MJ, Sedar AW, Mooers SU, Lefer AM: Lack of endothelium-dependent relaxation in coronary resistance arteries of cholesterol-fed rabbits. Am J Physiol 1989; 256:C591–C597.
72. Yamamoto H, Bossaller C, Cartwright J Jr, Henry PD: Videomicroscopic demonstration of defective cholinergic arteriolar vasodilation in atherosclerotic rabbit. J Clin Invest 1988; 81:1752–1758.
73. Vita JA, Treasure CB, Nabel EG, et al: Coronary vasomotor response to acetylcholine relates to risk factors for coronary artery disease. Circulation 1990; 81:494–497.
74. Casino PR, Kilcoyne CM, Quyyumi AA, Hoeg JM, Epstein SE, Panza JA: Role of nitric oxide in the endothelium-dependent vasodilation of hypercholesterolemic patients. Circulation 1992; 86:I-618.
75. Drexler H, Zeiher AM, Meinertz T, Just H: Correction of endothelial

dysfunction in coronary microcirculation of hypercholesterolemic patients by L-arginine. Lancet 1991; 338:1546–1550.

76. Creager MA, Cooke JP, Mendelsohn ME, et al: Impaired vasodilation of forearm resistance vessels in hypercholesterolemic humans. J Clin Invest 1990; 86:228–234.

77. Quyyumi AA, Dakak N, Andrews NP, et al: Nitric oxide activity in human coronary atherosclerosis. Circulation 1994; 90:1718 (Abstract).

78. Mügge A, Elwell JH, Peterson TE, Harrison DG: Release of intact endothelium-dependent relaxing factor depends on endothelial superoxide dismutase activity. Am J Physiol 1991; 260:C219–C225.

79. Minor RL Jr, Myers PR, Guerra R Jr, Bates JN, Harrison DG: Diet-induced atherosclerosis increases the release of nitrogen oxides from rabbit aorta. J Clin Invest 1990; 86:2109–2116.

80. Ohara Y, Peterson TE, Harrison DG: Hypercholesterolemia increases endothelial superoxide anion production. J Clin Invest 1993; 91:2546–2551.

81. Panza JA, Quyyumi AA, Brush JE Jr, Epstein SE: Abnormal endothelium-dependent vascular relaxation in patients with essential hypertension. N Engl J Med 1990; 323:22–27.

82. Panza JA, Casino PR, Kilcoyne CM, Quyyumi AA: Role of endothelium-derived nitric oxide in the abnormal endothelium-dependent vascular relaxation of patients with essential hypertension. Circulation 1993; 87:1468–1474.

83. Treasure CB, Klein JL, Vita JA, et al: Hypertension and left ventricular hypertrophy are associated with impaired endothelium-mediated relaxation in human coronary resistance vessels. Circulation 1993; 87:86–93.

84. Katz SD, Schwarz M, Yuen J, LeJemtel TH: Impaired acetylcholine-mediated vasodilation in patients with congestive heart failure. Circulation 1993; 88:55–61.

85. Egashira K, Inou T, Hirooka Y, Kai H, Sugimachi M, Suzuki S, Kuga T, Urabe Y, Takeshita A: Effects of age on endothelium-dependent vasodilation of resistance coronary artery by acetylcholine in humans. Circulation 1993; 88:77–81.

86. Yasue H, Matsuyama K, Koumura K, Morkiami Y, Ogawa H: Responses of angiographically normal human coronary arteries to intracoronary injection of acetylcholine by age and segment. Circulation 1990; 81:482–490.

87. Dakak N, Gilligan DM, Andrews NP, Diodati JG, Panza JA, Schenke WH, Cannon RO III, Quyyumi AA: Peripheral vascular endothelial dysfunction in patients with multiple coronary risk factors. Circulation 1993; 88:1618.

88. Quyyumi AA, Dakak N, Gilligan DM, Andrews NP, Diodati JG, Panza JA, Cannon RO III: Peripheral vascular endothelial dysfunction in syndrome X patients with endothelial dysfunction of the coronary microvasculature. Circulation 1993; 88:1369.

89. Quyyumi AA, Dakak N, Arora S, Gilligan DM, Panza JA, Andrews, NP, Cannon RO: Contribution of nitric oxide release to metabolic vasodilation in the human heart. J Am Coll Cardiol 1994; 1:274A.

90. Cox DA, Vita JA, Treasure CB, Fish RD, Alexander RW, Ganz OP, Selwyn AP: Atherosclerosis impairs flow-mediated dilatation of coronary arteries in humans. Circulation 1989; 80:458–465.

91. Folts JD, Crowell EB, Rowe GG: Platelet aggregation in partially obstructed vessels and its elimination with aspirin. Circulation 1976; 54:365–370.

92. Folts JD, Gallagher K, Rowe GG: Blood flow reductions in stenosed canine coronary arteries: vasospasm or platelet aggregation? Circulation 1982; 65:248–255.
93. Tofler GH, Brezinski DA, Schafer A, et al: Concurrent morning increase in platelet aggregability and the risk of myocardial infarction and sudden cardiac death. N Engl J Med 1987; 316:1514–1518.
94. Pohl U, Busse R: EDRF increases cyclic GMP in platelets during passage through the coronary vascular bed. Circulation Res 1989; 65:1798–1803.
95. Yao S-K, Ober JC, Krishnaswami A, Ferguson JJ, Anderson V, Golino P, Buja LM, Willerson JT: Endogenous nitric oxide protects against platelet aggregation and cyclic flow variations in stenosed and endothelium-injured arteries. Circulation 1992; 86:1302–1309.
96. Vanhoutte PM, Houston DS: Platelets, endothelium, and vasospasm. Circulation 1985; 72:728–734.
97. Andrews NP, Dakak N, Schenke WH, Quyyumi AA: Platelet-endothelial interactions in humans: changes in platelet cyclic guanosine monophosphate content in patients with endothelial dysfunction. Br Heart J 1994; 71:350 (Abstract).
98. Garg UC, Hassid A: Nitric oxide-generating vasodilators and 8-bromocyclic guanosine monophosphate inhibit mitogenesis and proliferation of cultured rat vascular smooth muscle cells. J Clin Invest 1989; 83:1774–1777.
99. Cayatte AJ, Palacino JJ, Horten K, Cohen RA: Chronic inhibition of nitric oxide production accelerates neointima formation and impairs endothelial function in hypercholesterolemic rabbits. Arterioscler Thromb 1994; 14:753–759.
100. Naruse K, Shimizu M, Muramatsu M, Toki Y, Miyazaki Y, Okumura K, Hashimoto H, Ito T: Long-term inhibition of NO synthesis promotes atherosclerosis in the hypercholesterolemic rabbit thoracic aorta. Arterioscler Thromb 1994; 14:746–752.
101. Diodati JG, Quyyumi AA, Hussain N, Keefer LK: Complexes of nitric oxide with nucleophiles as agents for the controlled biological release of nitric oxide: anti-platelet effect. Thromb Hem 1993; 70:654–658.
102. Diodati JG, Quyyumi AA, Hussain N, Keefer LK: Complexes of nitric oxide with nucleophiles as agents for the controlled biologic release of nitric oxide: hemodynamic effect. J Cardiovasc Pharmacol; 1993; 22:287–292
103. Creager MA, Gallagher SJ, Girerd XJ, Coleman SM, Dzau VJ, Cooke JP: L-arginine improves endothelium-dependent vasodilation in hypercholesterolemic humans. J Clin Invest 1992; 90:1248–1253.
104. Panza JA, Casino PR, Badar DM, Quyyumi AA: Effect of increased availability of endothelium-derived nitric oxide precursor on endothelium-dependent vascular relaxation in normals and in patients with essential hypertension. Circulation 1993; 87:1475–1481.
105. Casino PR, Kilcoyne CM, Quyyumi AA, Hoeg JM, Panza JA: Investigation of decreased availability of nitric oxide precursor as the mechanism responsible for impaired endothelium-dependent vasodilation in hypercholesterolemic patients. J Am Coll Cardiol 1994; 23:844–850.
106. Cooke JP, Andon NA, Girerd XJ, Hirsch AT, Creager MA: Arginine restores cholinergic relaxation of hypercholesterolemic rabbit thoracic aorta. Circulation 1991; 83:1057–1062.

107. Vanhoutte PM, Boulanger CM, Illiano SC, Nagao T, Vidal M, Mombouli J-V: Endothelium-dependent effects of converting-enzyme inhibitors. J Cardiov Pharmacol 1993; 22:S10–S16.

108. Mügge A, Elwell JH, Peterson TE, Hofmayer TG, Heistad DD, Harrison DG: Chronic treatment with polyethylene-glycolated superoxide dismutase partially restores endothelium-dependent vascular relaxations in cholesterol-fed rabbits. Circ Res 1991; 69:1293–1300.

109. Scandinavian Simvastatin Survival Study Group: Randomized trial of cholesterol lowering in 4444 patients with coronary heart disease: the Scandinavian Simvastatin Survival Study (4S). Lancet 1994; 344:1383–1389.

110. Harrison DG, Armstrong ML, Freiman PC, Heistad DD: Restoration of endothelium-dependent relaxation by dietary treatment of atherosclerosis. J Clin Invest 1987; 80:1808–1811.

111. Egashira K, Hirooka Y, Kai H, Sugimachi M, Suzuki S, Inou T, Takeshita A: Reduction in serum cholesterol with pravastatin improves endothelium-dependent coronary vasomotion in patients with hypercholesterolemia. Circulation 1994; 89:2519–2524.

112. Lin SG, Flaherty JT: Cross-over from intravenous to transdermal nitroglycerin therapy in unstable angina pectoris. Am J Cardiol 1985; 56:742–747.

113. Hackett D, Davies G, Chierchia S, Maseri A: Intermittent coronary occlusion in acute myocardial infarction: value of combined thrombolytic and vasodilator therapy. N Engl J Med 1987; 317:1055–1059.

114. Maseri A, L'Abbate A, Baroldi G, et al: Coronary spasm as a possible cause of myocardial infarction: a conclusion derived from the study of 'pre-infarction' angina. N Engl J Med 1978; 299:1271–1277.

115. Cipriano PR, Koch FH, Rosenthal SJ, Baim DS, Ginsburg R, Schroeder JS: Myocardial infarction in patients with coronary artery spasm demonstrated by angiography. Am Heart J 1983; 105:542–547.

116. Roberts WB, Curry RC, Isner JM, et al: Sudden death in Prinzmetal's angina with coronary spasm documented angiographically. Am J Cardiol 1982; 50:203–210.

117. Nademanee K, Intarachot V, Josephson MA, Singh BH: Circadian variation in occurrence of transient overt and silent myocardial ischemia in chronic stable angina and comparison with Prinzmetal's angina in men. Am J Cardiol 1987; 60:494–498.

4

Circadian Variations in Electrophysiological Parameters and Cardiac Repolarization

Xavier Viñolas, MD, Angel Moya, MD,
Juan Cinca, MD, Antonio Bayés de Luna, MD

Introduction

As with many other systems in the human body, the cardiovascular system is subjected to circadian rhythm. Daily variations in heart rate and arterial blood pressure have been demonstrated.[1-5] A circadian pattern in the incidence of stable[6,7] and unstable angina[7-9] has also been described, with a marked increase in ischemic episodes between 7 AM to 10 AM. Circadian variations in the pattern of presentation of ventricular[10-12] and supraventricular[13] cardiac arrhythmias have also been shown, both in untreated and in treated patients.[14] Whether these changes can be attributed only to circadian variations in sympathetic-parasympathetic tone or to daily activity[1,2,15-17] or positional changes[18] is not clear.

While the variations in the occurrence of malignant ventricular arrhythmias can be attributed, at least partially, to the circadian changes in the presentation of ischemia,[19] the most frequent supraventricular arrhythmias, such as atrioventricular (AV) nodal reentrant tachycardias

From: Deedwania PC (ed): *Circadian Rhythms of Cardiovascular Disorders.* ©Futura Publishing Co., Inc., Armonk, NY, 1997.

and AV reciprocant tachycardias, occur usually in young people without heart disease. The circadian pattern observed in the clinical presentation of these arrhythmias[13] can be due to changes in the electrophysiological properties of the different limbs that participate in the triggering and in the maintenance of these arrhythmias. Those structures are the atria, the AV node, the ventricles, and the AV accessory pathway. Thus, the analysis of the circadian changes of the electrophysiological properties of these structures can help in understanding the possible variations in the clinical presentation of these arrhythmias.

Circadian Variation in Electrophysiological Parameters

Methods for Circadian Measurements of Electrophysiological Parameters

To assess the presence of circadian variations in the electrophysiological properties of the different structures of the heart, the possibility of stable and prolonged intracavitary recordings and pacing is important.

Conventional diagnostic electrophysiological studies include the measurement of AV conduction intervals (AH and HV), the calculation of sinus node recovery time, and the determination of refractory periods of atria, AV node, ventricle and, if present, the accessory pathway. The study must also look for the possible presence of dual AV nodal pathways or AV accessory pathways and, when present, supraventricular tachycardia must be induced with programmed extrastimulation (PES) from the atria or from the ventricle. In order to achieve these objectives, multiple intracavitary tetrapolar electrocatheters must be used. One must be placed at the right ventricular apex for pacing the right ventricle; another must be positioned in the tricuspid annulus near the AV node region in order to record the His bundle potential; and a third electrocatheter must be placed at the right atrium. In patients who may have an atrioventricular accessory pathway, a fourth electrocatheter must be introduced in the coronary sinus to look for the presence of left atrioventricular connections. The catheters placed in the right atrium and in the AV node region are not stable enough to remain in place for a prolonged period of time. Additionally, the displacement of the His bundle catheter can cause dangerous ventricular arrhythmias.

In order to allow a stable recording of His bundle potential during prolonged monitoring, we designed an octapolar catheter.[20] This catheter had a distal pair of electrodes with an interelectrode distance of 1 cm. The second and the third pair of electrodes within an interelectrode distance of 0.5 cm were located 6 and 7 cm, respectively, from the catheter tip. The fourth pair of electrodes, with an interelectrode distance of 1 cm, was located 12 cm from the tip. The aim was to be able to stimulate the apex of the right ventricle with the distal pair of electrodes and to record the His bundle activity with the second or third pairs and the right atrial activity with the fourth pair with only one catheter.

The catheter was tested in 45 consecutive patients. In 39 of the 45 (86%) patients in whom the catheter was placed at the right ventricular apex, a His bundle potential of a good quality was recorded in the baseline study. In 11 of those 39 patients, the catheter was left in place, obtaining stable intracavitary recordings during a 24-hour period.[20]

In 38 patients, an eletrophysiological study was indicated because of episodes of paroxysmal supraventricular tachycardia (n=25) or syncope of unknown origin (n=13).[21,22]

In all of these patients, a baseline electrophysiological study was performed between 11 AM and 2 PM. In this first baseline study, three electrocatheters were introduced. An octapolar electrocatheter was advanced through the right femoral vein and was positioned at the right ventricular apex, the His bundle potential was recorded with the second or third pair of electrodes, and the right atrial activity was recorded with the proximal pair. Two tetrapolar electrocatheters were additionally introduced: one of them through the right femoral vein was positioned at the high right atrium and the other one through the left subclavian vein was introduced to the coronary sinus. The baseline study included the determination of the RR interval, the measurement of intracavitary AV conduction intervals (AH and HV), and the calculation of sinus node recovery time. Programmed atrial and ventricular electrical stimulation was performed to assess the effective refractory period (ERP) of the atria, AV node, and right ventricle. The possible presence of dual AV nodal pathways or an AV accessory pathway was analyzed. In those patients with an accessory pathway, the ERP of the anterograde and retrograde conduction through the accessory pathway was determined. In patients with either an accessory pathway or dual AV nodal pathways, the inducibility of reciprocant tachycardia with programmed electrical stimulation was assessed. Once the baseline study was finished, the catheter placed at the high right atrium was withdrawn, whereas the octapolar catheter placed in the right apex and the catheter located in the coronary sinus were left in place and sutured to the skin to minimize electrode dislodgment. The patients were transferred to a diurnal lightened room. With these two catheters we were able to record the right atrial and His bundle activity and to stimulate the right ventricle and left atrium from the coronary sinus throughout

24 hours. Every hour, RR, AH, and HV intervals were measured during sinus rhythm as well as during fixed atrial pacing. In addition, at intervals of 2 hours, the sinus node recovery time and refractory periods were determined. The episodes of induced tachycardia related to stimulation attempts were analyzed.[23] The programmable stimulator and the recording system were placed outside the patient's room to reduce the interference of the study protocol with the sleep periods. After 24 hours, the catheters were withdrawn and the patients were discharged from the hospital.

The ability to electrically induce supraventricular tachycardia was expressed as the percentage of programmed electrical stimulation inducing at least one episode of supraventricular tachycardia. The significance of arrhythmia inducibility was evaluated by x^2 analysis for trend. A $p<0.05$ was considered significant.

An average of nine (range 7–13) bedside electrophysiological studies were performed in each patient without complications.

Variations of Refractory Periods in Successive EPS Studies

In the baseline electrophysiological study, an atrioventricular accessory pathway was present in 19 patients, eight with bidirectional conduction, and 11 with only retrograde conduction; in 16 cases the accessory pathway was left-sided and in three, right-sided. In six patients, there was evidence of dual AV nodal pathway (Table 1).

Analysis of variance for the sequential changes of ERP of the atria, AV node, and right ventricle showed a statistically significant variation. Compared with the first electrophysiological study, during midnight and early morning there was a significant prolongation of the ERP of the atria ($p<0.001$), AV node ($p=0.002$, Fig. 1) and right ventricle ($p<0.001$), preceded by a transient shortening of these parameters from 7 to 8 PM and returning to baseline values in the last study late in the morning. The sinus node rate, expressed as RR interval, fol-

Table 1

Electrophysiological Findings in 38 Patients

Bidirectionnal left-sided Kent bundle	5
Bidirectionnal right-sided Kent bundle	3
Concealed left-sided Kent bundle	11
Dual AV nodal pathways	6
Normal electrophysiological study	13

AV NODAL ERP
msec.

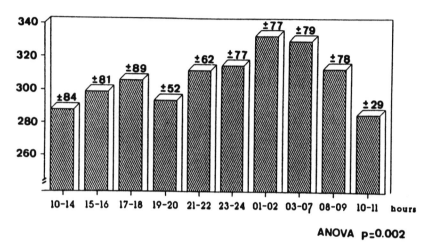

Figure 1: Sequential values of the AV nodal effective refractory period (ERP) throughout the study period.

lowed a significant prolongation at midnight, paralleling the changes in cardiac refractoriness. The sinus node recovery time also showed a significant lengthening from 12 and 3 AM, returning to baseline values the following morning. AV conduction intervals AH and HV did not show any significant variation during the period of study.

In patients with AV accessory pathways, sequential measurements evidenced a significant prolongation of retrograde Kent bundle ERP (p<0.005) at midnight and early morning preceded by a transient shortening from 7 and 8 PM (Fig. 2). Refractoriness of antegrade conduction of Kent bundle, present in only eight patients, tended to prolong at midnight, but this change was not statistically significant (Fig. 3). In six patients with dual AV nodal pathway, the ERP of the fast pathway experienced a lengthening at midnight and early morning preceded by a brief shortening from 7 to 8 PM. In contrast, the slow AV pathway showed a different daily variation. In five patients, two or more electrophysiological tests showed conduction block through the slow nodal pathway. This phenomenon was not related to the nocturnal part of the day but it tended to occur when ERP of the fast AV nodal pathway showed a concurrent shortening.

RETROGRADE KENT BUNDLE ERP
msec.

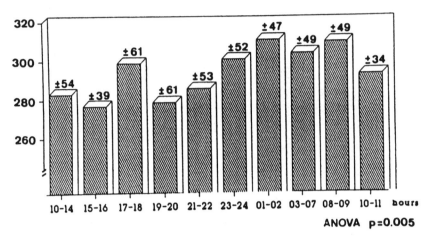

Figure 2: Sequential values of the effective refractory period (ERP) of the retrograde conduction through the Kent bundle during the study period. (Used with permission from Cinca J, et al.[21])

ANTEGRADE KENT BUNDLE ERP
msec.

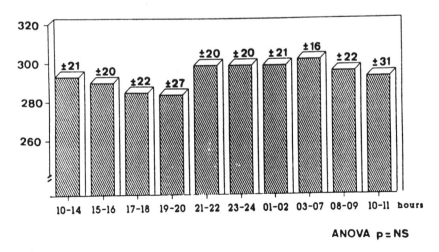

Figure 3: Sequential values of effective refractory period (ERP) of the antegrade conduction through the Kent bundle during the study period.

Variations in Tachycardia Induction

In 17 out of the 19 patients with Kent bundles, reciprocating tachycardia involving the accessory pathway was induced by programmed electrical stimulation either from the apex of the right ventricle or from the coronary sinus. The ability to induce tachycardia varied throughout the study protocol. The percentage of PES that induced arrhythmia progressively decreased through the afternoon, showed a transient increase from 9 and 10 PM, decreased significantly at midnight and early morning, and tended to recover during the ensuing morning. These changes were observed either from the right ventricle or from the coronary sinus (Fig. 4). Although not significant, there was a trend to observe a prolongation of the tachycardia cycle length at midnight whenever the arrhythmia was inducible. In only three out of six patients with dual AV nodal pathways, a reentrant tachycardia was inducible at the baseline electrophysiological study. In two of them, the tachycardia was no longer inducible at midnight, and in the remaining patient, the induced tachycardia had a longer cycle length at midnight than at the baseline study.

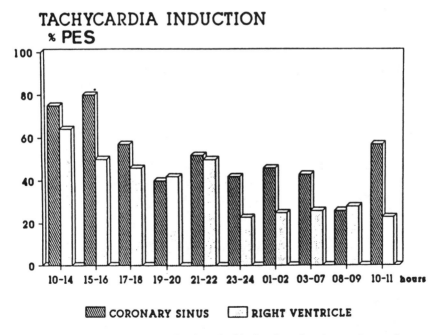

Figure 4: Sequential changes in electrical induction of reciprocating tachycardia during the study period.

These results show that effective refractory periods of cardiac structures that participate in the triggering and maintenance of supraventricular reentrant tachyarrhythmias have a well-defined circadian pattern with a significant prolongation at midnight and early morning, with a brief shortening from 8 and 10 PM, closely preceding the midnight prolongation and with a return to baseline values in the ensuing morning. These changes run parallel with the circadian variation observed in the ability to induce tachycardia, either in patients with AV accessory pathway or in patients with dual AV nodal pathways, with a diminished capacity at midnight and early morning and a brief enhancement from 8 to 10 PM.

Due to the characteristics of the study, whereby the patients were admitted to a lightened room and were in bed rest, there is a low probability that the changes observed were due to modifications in body position or daily activity. The most probable cause of these changes is the circadian modification of the sympathetic-parasympathetic tone that is present throughout the day.[24,25] The changes observed in spontaneous sinus rhythm in these patients suggest that the normal circadian pattern is preserved in this population.

These data give a physiopathological basis to the observation that spontaneous clinical episodes of supraventricular reentrant tachycardia have a circadian pattern, with a maximal incidence at 6–8 PM and a corresponding minimum at 4 AM.[13] According to our study, circadian variations of ERP of the structures that participate in the genesis and maintenance of reciprocant supraventricular tachycardias can play a critical role at the clinical appearance of these tachycardias, regardless of the changes in the body position or in the daily activity. Even though radiofrequency catheter ablation is currently the first-line treatment for most of these patients, antiarrhythmic drugs can be used in patients who refuse ablative therapy. In those patients, drug therapy can be concentrated at the less protected parts of the day, i.e., late morning and late evening.[26] This can help to increase affectivity, reducing the total dose and, consequently, decreasing the likelihood of secondary effects.

Circadian Variations in Cardiac Repolarization

Introduction

We have already seen that invasive electrophysiological parameters vary during the day and can show a circadian pattern. It is also well known that both cardiac death and sudden cardiac death show a circadian pattern.[27-30] In cases in which the ventricular arrhythmia is not linked to an acute ischemic episode, other triggering factors (disturbance of the sympathetic-vagal balance, increased malignant ventricular premature beats, ischemic episodes, etc.), act on a vulnerable myocardium (postinfarction scar, left ventricular hypertrophy, etc.) and induce a malignant ventricular arrhythmia. A vulnerable myocardium may also exist for days, months, or years with no complication. The occurrence of arrhythmia at some time in the follow-up depends on the appearance of a specific trigger or on the interaction of various triggering factors.

Invasive electrophysiological parameters are very important in order to increase our knowledge of the fine underlying mechanisms, but they have no clinical value for application in these large populations, for example postinfarction patients, to evaluate risk of arrhythmic events. Therefore, it is important to find noninvasive markers reflecting electrophysiological changes in order to be able to evaluate these large populations.

The patients' prognoses, especially those of postinfarction patients, depend on three main factors: ischemia, electrical instability, and left ventricular dysfunction. Malignant ventricular arrhythmia is the maximum expression of electrical instability. These three factors interact, and disturbance or modification of any one of these factors may lead to changes in the other two. However, the purpose of this chapter is not to describe in detail the interactions between these parameters.

Currently, there are many parameters involved in evaluating electrical instability: ventricular late potentials, programmed electrical stimulation, ECG disturbances, arrhythmias on Holter recordings, disturbances in the repolarization parameters, etc., and RR variability as an expression of alterations of the autonomic nervous system. Recently, Rosembaun et al.[31] have shown that patients with event subtle alternans of ST-T wave have an increased risk of arrhythmic events during follow-up. QT dispersion in surface ECG is another measure of cardiac repolarization and theoretically could reflect refractory dispersion and may favor ventricular arrhythmias. It has also been reported[32,33] that patients with greater QT dispersion are at higher risk for ventricular arrhythmias or sudden death during follow-up.

Study of the circadian rhythms of these abnormalities is a new approach for evaluating the importance of triggering factors in these patients. Table 2 shows the different parameters and techniques used to stratify risk in postmyocardial infarction patients. We will evaluate in this chapter only the dynamic behavior of repolarization parameters, especially the behavior of the QT interval, an area of great interest for our group. We will review the behavior of the QT interval in healthy

Table 2

Parameters and Techniques Used to Stratify Risk in Postmyocardial Infarction Patients

	Parameters	Techniques
Electrical instability	Ventricular arrhythmias	Holter PES
	Autonomic nervous system •Heart rate variability •QT interval	Holter
	Anatomic substrate	Late potentials (conventional or Holter techniques)
Ischemia	Coronary flow	Exercise testing Holter Imaging techniques
	Coronary stenosis	Coronary angiography
	Prethrombotic state	Blood test
Left ventricular dysfunction	Diastolic function	Echo-Doppler
	Systolic function	Echocardiography and angiography and isotopic techniques
	State of RAA axis	Blood test

PES= Programmed electrical stimulation; MNR= Magnetic nuclear resonance; RAA= renin-angiotensin.

subjects and in postmyocardial infarction patients with and without ventricular arrhythmias.

QT Interval and Malignant Ventricular Arrhythmias

The relationship between the duration of the QT interval and presence of malignant ventricular arrhythmias is known and its maximum expression is the presence of congenital long QT syndrome. In some cases, these patients have episodes of "torsade de pointes" ventricular tachycardia, which can lead to sudden death.

The interest in the QT interval also derives from observations that excessive prolongation of the QT interval with group I drugs is associated in some cases with proarrhythmia. On the other hand, prolongation of the QT interval by amiodarone, within limits that have not been well established, is considered a good parameter of the drug's effectiveness.

The QT interval is a simple measurement on the surface electrocardiogram. Thus, the possibility of finding a relationship between this measurement and the presence of malignant ventricular arrhythmias would be very important. However, measurement of the QT interval always has limitations. It is a simple measurement of cardiac repolarization, which is an extremely complex process that is influenced by many factors: presence or absence of underlying organic heart disease, the autonomic nervous system, circulating catecholamines, electrolytes, drugs (cardiac and noncardiac), and others. Moreover, the QT interval varies with heart rate, being shorter in shorter cycles, an effect that is usually corrected by the Bazett formula,[34] which somewhat distorts the results. This distortion of results occurs mainly at the upper and lower limits of heart rate. However, heart rate is not the only factor that changes QT; for instance, at the same heart rate, isoproterenol infusion shortens the QT interval. Likewise, in patients with VVI pacemakers with a fixed heart rate, the QT interval becomes shorter with exercise. This indicates that the QT interval is influenced by a number of factors, many of them dependent on the autonomic nervous system, which will not be discussed extensively. Therefore, the study of the dynamic behavior and circadian rhythm of the QT interval is very important because it reflects this multiple interaction. Circadian variations in the QT interval examined in relation to the circadian rhythm of malignant ventricular arrhythmias or sudden death may clarify some of the triggering mechanisms of ventricular arrhythmias and sudden cardiac death.

QT Variations in Healthy Subjects

The QT interval reflects the duration of depolarization and repolarization. As repolarization is much more prolonged and variations in depolarization (except for the appearance of bundle branch block) are less variable, changes in the QT interval essentially reflect changes in repolarization. The QT interval becomes shorter as the cardiac cycle becomes shorter. Correction of the QT interval using the Bazett formula is known to be inexact and there are other correction formulas,[35] the most exact of them apparently being the normograms (based on analysis of the QT/RR ratio in each group) that show by how many milliseconds QT must be corrected for each heart rate.[36] The correction for heart rate is an important tool that allows changes in the QT interval to be evaluated with some, but not all, independence of changes in heart rate. The simplicity of the Bazett formula explains why, although it is not exact, it is used almost universally.

We studied the behavior of the QTc interval on the Holter tapes of 10 healthy persons compared to those of postinfarction patients with and without malignant ventricular arrhythmia.[37,38] The QTc interval was more prolonged between 11 PM and 11 AM than during the other 12 hours of the day (Fig. 5). The mean QT interval was shorter (in the

CIRCADIAN RHYTHM OF QTc

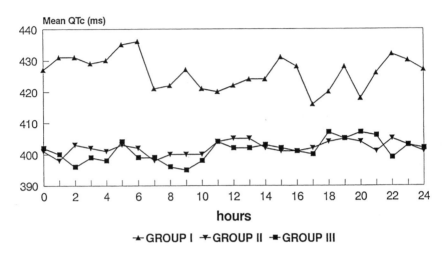

Figure 5: Mean hourly QTc in different groups of patients. Group I corresponds to postinfarction patients with ventricular arrhythmias during follow-up. Group II corresponds to postinfarction patients without ventricular arrhythmias during follow-up. Group III corresponds to normal subjects. We can see that circadian variations of QTc are similar in normal subjects and postinfarction patients without ventricular arrhythmias. Mean hourly QTc is longer all day in postinfarction patients with malignant ventricular arrhythmias, especially during the night. (Used with permission from Cinca J, et al.[21])

healthy subjects) than in the postinfarction patients who had malignant ventricular arrhythmia, but was similar to that of postinfarction patients without malignant ventricular arrhythmia (Table 3). Ninety percent of the healthy subjects had QT intervals over 440 ms at some time during the day, but none of the tapes analyzed showed peak QTc intervals in excess of 500 ms so this value should be considered as the cutoff point for normality (Table 4).

Vervaet et al.[39] studied variations in the QT interval on the surface ECGs of 84 healthy subjects and also found a daily rhythm in QTc interval measurements. They analyzed only ECGs recorded at different times of the day between 8 AM and 4 PM. This restricts the possibility of studying strictly circadian behavior, but variations in the QT interval were found in spite of these limitations and the short time period. The minimum QTc value was found between 10 AM and 12 AM.

Murakawa et al.[40] analyzed QTc interval values in different popu-

Table 3

Automatic QTc Analysis in the Three Studied Groups

Groups	Group I (n=14)	Group II (n=28)	Group III (n=10)	p value I vs. II	p value I vs. III
Total of beats automatically analyzed	682,960	1,276,498	563,910	—	—
Global QTc	425±15	408±19	402±20	<0.005	<0.001
Total number of peaks of QTc >500 msec	11,114 (1.62%)	823 (0.06%)	0 (0%)	<0.005	<0.005
Patients with peaks of QTc >500 msec	7 (50%)	2 (7%)	none	<0.005	<0.03
Patients with clusters of peaks of QTc >500 msec	4 (28%)	none	none	<0.02	<0.02

Group I: Postmyocardial infarction patients who presented a secondary life-threatening arrhythmia; *Group II:* postmyocardial infarction patients who did not present life-threatening arrhythmias; *Group III:* healthy subjects.

Table 4

Healthy Subjects with Peaks of QTc Lengthening Measured Automatically in Holter Recordings According to a Determined Cutoff of QTc Value

Peaks of QTc Interval (msec)	n=10
>440	9 (90%)
>460	4 (40%)
>480	3 (30%)
>500	0 (0%)

lation groups (healthy subjects, postmyocardial infarction patients, and diabetic neuropathy patients). They analyzed manually sets of six QT intervals every 10 minutes throughout the day and night. Although abrupt transitory changes in the QT interval (QT peaks) may pass undetected with this method, in contrast with automatic analysis, a sufficient number of QT intervals is studied to permit the circadian rhythm of the QT interval to be evaluated. Instead of using the Bazett correction formula, the Murakawa study analyzed the regression of QT/RR using three different formulas. In the group of 23 healthy subjects, the nocturnal QTc interval was found to be longer than the diurnal QTc interval (400±20 ms versus 384±22 ms). This prolongation of the QT interval coincided with periods of greater bradycardia, that is, with longer RR intervals. They also found less variation and variability in the QTc intervals of patients with ischemic heart disease and of patients with diabetes mellitus. The greater variability of the QT interval is similar to the greater variability of heart rate in healthy subjects compared to patients after myocardial infarction or in patients with heart disease.

Rasmusen et al.[41] analyzed variations in the QT interval in relation to time of day in 60 healthy subjects, measuring QT intervals each hour. Although they did not use automatic analysis, which allows evaluation of all QT intervals and of transitory changes in this interval, they obtained a basic idea of QT interval variations throughout the day. They found that the QTc interval, corrected for heart rate using the Bazett formula or a regression equation, was longer in the measurements made between 2 AM and 6 AM. When the factors that independently influence QT interval were analyzed, correlations were found with heart rate, naturally, and with the time of day at which the ECG was recorded.

Ong et al.[42] compared the circadian variations in QT interval, QTc, and heart rate in patients with diabetic neuropathy and healthy subjects. In a group of 13 healthy subjects, they observed the longest QT interval between midnight and 6 AM, a period during which heart rate also was lower, suggesting vagal predominance during this period. The decreased RR variability, higher heart rate, and shorter QTc interval in patients with diabetic neuropathy suggest disturbances in vegetative tone in the form of increased sympathetic tone (or decreased vagal tone), which could trigger electrical instability and provoke malignant ventricular arrhythmia.

Sarma et al.[43] analyzed automatically the circadian variation in a small group of healthy subjects (n=6) compared to the circadian variation found in patients with hypothyroidism before and after tiroxine treatment. QT was normalized for heart rate, using an exponential formula. The circadian variation of QT was analyzed using a mean QTc of 3-hour period. They found out that circadian variation of QTc was present in normal subjects, with a longer QTc during night time, when heart rate is also lower. Interestingly, QTc remains prolonged in patients with hypothyroidism even 8 to 12 weeks after normalization of biochemical indexes.

It is therefore evident that there is circadian variation in the QTc interval of healthy subjects, although these changes usually are linked to changes in heart rate, with longer QT and QTc intervals occurring during periods of bradycardia. However, the fact that these abnormalities are always found regardless of the formula used to correct the QT interval indicates that these changes are to some degree independent of heart rate. As we have already explained, many factors influence the QT interval: sympathetic-vagal balance, circulating catecholamines, and others. Circadian variations in the repolarization parameters are therefore a physiological phenomenon. It is important to evaluate the increase or, perhaps, disappearance of circadian variations, for example, in groups of patients who die suddenly, for the purpose of better understanding the physiological phenomena that trigger sudden death. The study of circadian variations in the QT-regulating factors using noninvasive methods (e.g., evaluation of changes in RR variability that reflect sympathetic-vagal status) should clarify these variations.

QT Variations in Postinfarction Patients with and without Malignant Ventricular Arrhythmias

As we mentioned earlier, sudden cardiac death is the most important cause of death in the adult population of developed countries. The most important cause of sudden death in postinfarction patients is malignant ventricular arrhythmia (ventricular tachycardia, ventricular fibrillation), generally not associated with previous ischemic episodes.

In recent years, interest in the relationship between QT interval and postinfarction cardiac sudden death has grown. Various studies[44–51] have analyzed the relationship between the QT interval and the presence of malignant ventricular arrhythmias, all using surface ECG. This has important drawbacks because it is only an instantaneous determination in a dynamic process and may be responsible for discrepancies in the results of different studies. Thus, some studies have shown that prolongation of the QT interval is more important than determination of its value at a given moment because, as noted, the behavior of the QT interval in healthy persons depends on the time of day at which the ECG was recorded.

To determine if this hypothesis is correct, we analyzed the value of QT variations in the Holter tapes of postinfarction patients with and without malignant ventricular arrhythmias. A manual measurement was made[52] of the QT intervals on the Holter tape by selecting several beats per hour. These measurements do not provide information on subtle and transitory changes in the QT/QTc interval. We found no differences in mean QTc value between postinfarction patents with and without malignant ventricular arrhythmias. However, patients with malignant ventricular arrhythmias during follow-up had QTc peaks >500 ms more frequently than patients without these arrhythmic complications during follow-up.

Later we developed an algorithm for automatic measurement of the QT interval.[53] Two groups of postinfarction patients were studied. Group 1 consisted of 14 consecutive postinfarction patients admitted to the coronary unit of our hospital for ventricular tachycardia (eight patients) or after an aborted sudden death episode (six patients). Arrhythmia secondary to an acute ischemic episode was excluded in all patients. The second group included 28 postinfarction patients "matched" by clinical data, ejection fraction, and infarction site who did not have malignant ventricular arrhythmias in a similar period of time. There were no significant differences in the clinical characteristics of the two groups of patients (Table 5).

The mathematical algorithm for automatic measurement of the QT interval has been described in detail elsewhere.[53] It is important to emphasize that we analyzed the results based on a QT interval reaching the end of the T wave. Others use a QT interval ending at the peak of the T wave (peak QT) because it is considered simpler and more stable. However, use of peak QT values causes part of the information to be lost because QT may be prolonged exclusively at the expense of the terminal parts of the T wave. The QT interval was corrected using the Bazett formula, which is simple and widely used in spite of its previously mentioned limitations, and because it is the most widely accepted correction method. The data of the different QTc intervals were stored in digital form and represented as a trend that allowed exact assessment of the behavior of the QTc interval throughout the day in a practical form. In all, each point on the graph of the trend did not represent an isolated value of a QT interval but the mean of the QT intervals measured for 6 seconds.

Afterwards the algorithm was validated, comparing manual measurements with those done automatically.[53] The differences between automatic and manual measurements were similar to those of manual measurements by two experts. This validated the use of the method for

---

Table 5

Clinical Characteristics of Postmyocardial Infarction Patients with and without Life-Threatening Arrhythmias

	Group I (n=14)	Group II (n=28)	p value
Sex (M/F)	12/2	25/3	NS
Age (years)	59±13	57±10	NS
Anterior MI	9 (64%)	16 (57%)	NS
LV ejection fraction (%)	40±6	44±8	NS
LV ejection fraction <40%	6 (42%)	13 (46%)	NS
Angina	3 (21%)	7 (25%)	NS
Diabetes mellitus	3 (21%)	8 (28%)	NS
Hypertension	8 (57%)	15 (53%)	NS

Group I: Postmyocardial infarction patients with secondary life-threatening arrhythmia; *Group II:* postmyocardial infarction patients without life-threatening arrhythmias. MI = myocardial infarction. NS = not significant

analysis of larger population groups and, above all, the analysis of all QT values, which cannot be done manually. Automatic measurement is the only valid means of evaluating transitory changes in the value of QT intervals, and thus analyzing "peaks" of QT.

We analyzed mean QTc value, peaks of QTc (QTc values above a certain value), and clusters (groups of peak QT values lasting more that 1 minute).[37,38] The mean QTc interval was longer in patients with malignant ventricular arrhythmias than in patients without these arrhythmias: 425±15 ms versus 408±19 ms (Table 3).

The behavior of the QT interval in relation to time of day is shown in Figure 5. This figure confirms the tendency toward longer QTc values from 11 PM to 11 AM than from 11 AM to 11 PM (430±18 ms versus 425±19 ms). These differences were not statistically significant, perhaps because of the small sample size, but they suggested a trend and concurred with findings in healthy subjects.

QT peaks were analyzed for different cutoff points: >440, >460, >480, and >500 ms. Statistically significant differences were found only when the group of patients with QT interval values above 500 ms was analyzed (Table 6). Fifty percent of the postinfarction patients (group 1) who had malignant ventricular arrhythmias had QT values over 500 ms compared to only 7% (2 of 28) of postinfarction patients who did not have malignant ventricular arrhythmias (group 2). As we mentioned previously, none of the healthy subjects analyzed had peak QTc over 500 ms. When we examined the presence of clusters, we found that none of the postinfarction patients without malignant ventricular arrhythmias had clusters but 28% of the postinfarction patients with malignant ventricular arrhythmias did (Table 3). When the number of beats with QTc >500 ms was analyzed, postinfarction patients with malignant ventricular arrhythmias had long QTc in 1.62% of beats compared with only 0.06% of beats in postinfarction patients without

---

Table 6

Patients with Peaks of QTc Lengthening Measured Automatically in Holter Recordings According to a Determinated Cutoff of QTc Value

Peaks of QTc Interval (ms)	Group I (n=14)	Group II (n=28)	p value
>440	14 (100%)	20 (71%)	NS
>460	10 (71%)	14 (50%)	NS
>480	7 (50%)	8 (28%)	NS
>500	7 (50%)	2 (7%)	<0.005

Group I: Postmyocardial infarction patients who presented a secondary life-threatening arrhythmia; *Group II:* postmyocardial infarction patients who did not present life-threatening arrhythmias.

malignant ventricular arrhythmia. Figure 6 shows the graphic representation of postinfarction patients who had peak QTc >500 ms. It is important to note that these peaks sometimes occurred in groups and did not correspond to artifacts of automatic analysis.

As for the hourly distribution of QTc peaks, the number of QTc peaks per hour in postinfarction patients with malignant ventricular arrhythmias was 463±315 (Table 7). The trend in QTc intervals is shown in Figure 5.

QTc peaks >500 ms exhibited circadian rhythm, the percentage of QT peaks per hour being higher between 11 PM and 11 AM than during the rest of the day. The most relevant results are summarized in Table 7. This finding coincides with hourly mean QTc values, which were longer during the same hours of the day. The QTc clusters also had circadian behavior, with a higher incidence between 11 PM and 11 AM. If we analyze the total duration of QTc clusters that also occurred at periods that coincided with the periods in which patients are at "higher risk," QTc duration was longer in the same hours. The mean duration of clusters from 11 PM to 11 AM was 10.60±9 minutes and from 11 AM to 11 PM was 2.85±1.95 (Table 7). However, this behavior does not indicate whether the QT interval was a triggering factor in ma-

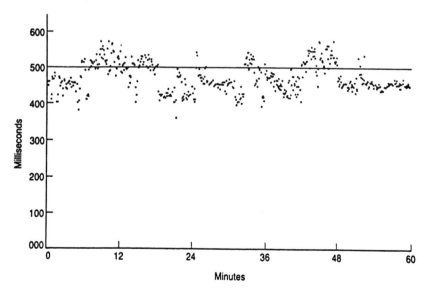

Figure 6: Representation of an hour of the trend of QTc interval in a postinfarction patient with a ventricular tachycardia. Note the presence of peaks of QTc >500 ms. We can see peaks of QTc greater than 500 ms and also grouped peaks ("clusters") at minute 10 and 48. (Used with permission from Cinca J, et al.[21])

Table 7

Circadian Rhythm of the Peaks and Clusters of QTc Lengthening Over 500 ms in Postmyocardial Infarction Patients with Malignant Ventricular Arrhythmias (Group I)

	24-hour Holter	11 AM– 11 PM	11 PM– 11 AM	p value
Number of peaks of QTc >500 ms per hour	463±315	336±176	590±375	<0.05
Percentage of peaks of QTc >500 ms per hour	4.17±2.83	3.02±1.58	5.31±3.37	<0.05
Number of grouped peaks (clusters) with QTc lengthening >500 ms per hour	1.29±1.27	0.84±0.83	1.75±1.48	NS
Mean duration (minutes) of the grouped peaks per hour	7.41±8.31	2.85±1.95	10.60±9.64	<0.04

Results represent the mean ±standard deviation taken from hourly Holter monitoring within the period of time indicated. The *p* value expresses the differences between time periods from 11 AM to 11 PM versus from 11 PM to 11 AM.

lignant ventricular arrhythmias or only an accompanying event. That is, the behavior of the QT/QTc interval varied throughout the day in postinfarction patients with malignant ventricular arrhythmias: there was a longer mean QTc/hour, larger number of QTc values >500 ms in absolute terms and as a percentage of beats, and more frequent and prolonged QTc clusters in the period between 11 PM and 11 AM.

Algra[54] recently published a study in which QTc interval was analyzed automatically on Holter tapes. In this author's study, prolongation of the QTc interval to more than 440 ms doubled the patients' risk of sudden death. Moreover, the presence of a short QTc interval (<400 ms) also was predictive of arrhythmic complications during follow-up. However, Algra's study is not entirely comparable to our study because only 50% of their highly heterogeneous group of patients were postinfarction patients and many of them were taking drugs that can modify QT interval.

We mentioned at the beginning of the chapter that the rate of occurrence of sudden death and malignant ventricular arrhythmias is greater between 6 AM and noon and we have seen the same circadian rhythm in QT value. Although this fact may be considered a link between changes in QT and sudden death, it does not demonstrate a cause-effect relationship. Prolongation of the QT interval may be a simple marker or a bystander that accompanies the true triggers of malignant ventricular arrhythmia.

As noted, the QT interval is influenced by sympathetic/parasympathetic tone. The QT/RR interval also varies with sympathetic or parasympathetic stimulation. Variations in the QT interval may coincide with disturbances in the autonomic nervous system that may trigger malignant ventricular arrhythmias.

We have seen that the presence of malignant ventricular arrhythmias and sudden cardiac death presents a circadian rhythm. Most of the actually known possible triggering factors such as ischemia, ventricular premature beats, etc., also show this circadian pattern. In the last few years, considerable attention has been focused on the important role of the autonomic nervous system as a triggering factor of malignant arrhythmias. Although we have not reviewed all of the repolarization parameters of the sympathetic-parasympathetic balance, some of them, such as the heart rate variability, also show this circadian rhythm. QT and QTc intervals are other important parameters for autonomic nervous system analysis, although they can be influenced by other factors (such as circulating catecholamines, drugs, etc.).

In this chapter, we have reviewed how QT interval presents a circadian rhythm in normal subjects. In postinfarction patients without malignant ventricular arrhythmias during follow-up, this circadian pattern remains but sudden prolongation of QTc values ("peaks") usually grouped in a cluster form appears. The circadian rhythm of these peaks is quite similar to the circadian pattern of sudden death. Nevertheless, this is not a proof that QTc prolongation is the triggering factor of the sudden death. This QTc prolongation may only be the manifestation of an underlying process such as a sudden imbalance of the autonomic nervous system. This circadian pattern of possible triggering factors opens a new way for sudden death mechanisms studies. Several mechanisms, triggers, modulators, etc., are already unknown and why an arrhythmia develops in a certain moment in the follow-up remains, in some cases, a "mystery."

A number of questions remain open: Does the circadian pattern of these parameters have any clinical relevance? Is it only an external manifestation of an underlying mechanism? Are there real trigger mechanisms? Is the loss of the circadian pattern a worse prognosis factor in postinfarction patients? Without any doubt, future technological improvements in Holter equipment, with the possibility of studying not only circadian variations of heart rate variability but also other repolarization parameters such as QT interval, T-wave alternans, etc., and their modification before an arrhythmic episode (using Holter recordings of AICD devices), will increase our knowledge of sudden death mechanisms.

References

1. Pieper C, Warren K, Peckering TG: A comparison of ambulatory blood pressure and heart rate at home and work on work and non-work days. J Hypertens 1993; 11:177–183.
2. Kohno I, Ishii H, Nakamura T, Tamura K: Relationship between activity levels and circadian blood pressure variations. Chronobiologia 1993; 20: 53–61.
3. Lund-Johansen P, Omvik P, White W: Measurement of long-term hemodynamic changes and the use of 24-hour blood pressure monitoring to evaluate treatment. Am J Cardiol 1994; 73(3):44A–49A.
4. Atkinson G, Witte K, Nold G, Lemmer B: Effects of age on circadian blood pressure and heart rate rhythms in patients with primary hypertension. Chronobiol Int 1994; 11:35–44.
5. Degaute JP, Van Cauter E, van de Borne P, Linkowski P: Twenty-four-hour blood pressure and heart rate in humans: a twin study. Hypertension 1994; 23:244–253.
6. Marchant B, Stevenson R, Vaishanv S, Wilkinson P, Ranjandayalan K, Timmis AD: Influence of the autonomic system on circadian patterns of myocardial ischemia: comparison of stable angina with the early postinfarction period. Br Heart J 1994; 71:329–333.
7. Parker JD, Testa MA, Jimenez AH, Tofler GH, Muller JE, Parker JO, Stone PH: Morning increase in ambulatory ischemia in patients with stable coronary artery disease: importance of physical activity and increased cardiac demand. Circulation 1994; 89:604–614.
8. Figueras J, Lidón RM: Circadian rhythm of angina in patients with unstable angina: relationship with extent of coronary disease, coronary reserve and ECG changes during pain. Eur Heart J 1994; 15:753–760.
9. Benhorin J, Banai SD, Moriel M, Gavish A, Keren A, Stern S, Tzivoni D: Circadian variations in ischemic threshold and their relation to the occurrence of ischemic episodes. Circulation 1993; 87:808–814.
10. Aronow WSW, Ahn C: Circadian variations of primary cardiac arrest or sudden cardiac death in patients aged 62 to 100 years (mean 82). Am J Cardiol 1993; 71:1455–1456.
11. Hohnloser SH, Klingenhen T: Insights into the pathogenesis of sudden cardiac death from analysis of circadian fluctuations of potential triggering factors. PACE 1994; 17:428–433.
12. Maron BJ, Kogan J, Proschan MA, Hecht GM, Roberts WC: Circadian variability in type of occurrence of sudden cardiac death in patients with hypertrophic cardiomyopathy. J Am Coll Cardiol 1994; 23:1405–1409.
13. Irwin JM, McCarthey EA, Wilkinson WE, Pritchett ELC: Circadian occurrence of symptomatic paroxysmal supraventricular tachycardias in untreated patients. Circulation 1988; 77:293–298.
14. Peters RW, Mitchell LB, Brooks MM, Echt DS, Barker AH, Capone R, Liebson PR, Greene HL: Circadian pattern of arrhythmic death in patients receiving encainide, flecainide or moricizine in the Cardiac Arrhythmia Suppression Trial (CAST). J Am Coll Cardiol 1994; 23:289–183.
15. Mulcahy D, Wright C, Sparrrow J, Cunningham D, Curcher D, Purcell H, Fox K: Heart rate and blood pressure consequences of an afternoon SIESTA

(Snooze-Induced Excitation of Sympathetic Triggered Activitiy) Am J Cardiol 1993; 71:611–614.

16. Bursztyn M, Mekler J, Watchel N, Ben Ishay D: Siesta and ambulatory blood pressure monitoring: comparability of the afternoon nap and night sleep. Am J Hypertens 1994; 7:217–221.

17. Van de Borne P, Nguyen H, Biston P, Linkowski P, Degaute JP: Effects of wake and sleep stages on the 24-h autonomic control of blood pressure and heart rate in recumbent men. Am J Physiol 1994; 266:H548–554.

18. Hiukuri HV, Niemela MJ, Ojala S, Rantala A, Ikaheimo MJ, Airaksinen KE: Circadian rhythms of frequency domain measures of heart rate variability in healthy subjects and patients with coronary artery disease: effects of arousal and upright posture. Circulation 1994; 90:121–126.

19. Hohnloser SH, Klingenheben T: Insights into the pathogenesis of sudden cardiac death from the analysis of circadian fluctuations of potential triggering factors. PACE 1994; 17:428–433.

20. Cinca J, Moya A, Bardají A, Figueras J, Rius J: Octapolar electrocatheter for His bundle recording and sequential bedside electrophysiologic testing. PACE 1988; 11:220–224.

21. Cinca J, Moya A, Bardají A, Rius J, Soler-Soler J: Circadian variations of electrical properties of the heart. Ann NY Acad Sci 1990; 601:222–233.

22. Cinca J, Moya A, Figueras J, Roma F: Circadian variations in the electrical properties of the human heart assessed by sequential bedside electrophysiologic testing. Am Heart J 1986; 112:315–321.

23. Cinca J, Moya A, Bardají A, Figueras JJ, Rius J: Daily variability of electrically induced reciprocating tachycardia in patients with atrioventricular accessory pathways. Am Heart J 1987; 114:327–333.

24. Hartikainen J, Tarkaianen I, Tahvanaien K, Mantysaari M, Lansimies E, Pyorala K: Circadian variation of cardiac autonomic regulation during 24-h bed rest. Clin Physiol 1993; 13:185–196.

25. Murakawa Y, Ajiki K, Usui M, Yamashita T, Oikawa N, Inoue H: Parasympathetic activity is a major modulator of the circadian variability of heart rate in healthy subjects and in patients with coronary artery disease or diabetes mellitus. Am Heart J 1993; 126:108–114.

26. Mulcahy D: Circadian variations in cardiovascular disease: implications for treatment. Br J Clin Pract 1994; 73:31–36 (Symp Suppl).

27. Muller JE, Stone PH, Turi ZG, and the MILIS Study Group: Circadian variation in the frequency of onset of acute myocardial infarction. N Engl J Med 1985; 313:1315.

28. Roberts R, Croft C, Gold HK, et al: Effects of propranolol on myocardial infarct size in a randomized, blinded, multicenter trial. N Engl J Med 1984; 311:218.

29. Muller JE, Ludmer PL, Willich SN, et al: Circadian variation in the frequency of sudden cardiac death. Circulation 1987; 75:131.

30. Willich SN, Levy D, Rocco MB, et al: Circadian variation in the incidence of sudden cardiac death in the Framingham Heart Study Population Study. Am J Cardiol 1987: 60:801–806.

31. Rosembaun D, Jackson L, Smith, et al: Electrical alternans and vulnerability to ventricular arrhythmias. N Engl J Med 1994; 330:235–241.

32. Zareba W, Moss AJ, Cessie L: Dispersion of ventricular repolarization and

arrhythmic cardiac death in coronary artery disease. Am J Cardiol 1994; 74(6):550–553.

33. Pye M, Quinn AC, Cobbe SM: QT interval dispersion: a non-invasive marker of susceptibility to arrhythmia in patients with sustained ventricular arrhythmias? Br Heart J 1994; 71:511–514.

34. Bazett HC: An analysis of the time-relation of the electrocardiogram. Heart 1920; 7:353–370.

35. Puddu PE, Jouve R, Mariotti S, et al: Evaluation of 10 QT prediction formula in 881 middle-aged men from seven country studies: emphasis on the cubic root Fridericia's equation. J Electrocardiol 1988; 21:219–229.

36. Karjalainen J, Viitasalo M, Mänttäri M, Manninen V: Relation between QT intervals and heart rates from 40 to 120 beats/min in rest electrocardiograms of men and a simple method to adjust QT interval values. J Am Coll Cardiol 1994; 23:1547–1553.

37. Homs E, Viñolas X, Guindo J, Martí V, Guri O, Bayés de Luna A: Automatic QTc measurement in Holter ECG as a marker of life-threatening ventricular arrhythmias in postmyocardial infarction patients. JACC 1993; 21:274A.

38. Homs E, Martí V, Laguna P, Viñolas X, Guindo J, Caminal P, Bayés de Luna A: Dynamic behavior of QTc measured in Holter tapes in different groups of postmyocardial infarction patients: evidence of circadian variation of QTc lengthening. JACC 1994;1A–484A.

39. Vervaet P, Amery W: Reproducibility of QTc measurements in healthy volunteers. Acta Cardiol 1993; 48:555–564.

40. Murakawa Y, Inoue H, Nozaki A, Sugimoto T: Role of sympathovagal interaction in diurnal variation of QT interval. Am J Cardiol 1992; 69:339–343.

41. Rasmusen V, Jensen G, Hansen JF: QT interval in 24-hour ambulatory ECG recordings from 60 healthy adult subjects. J Electrocardiol 1991; 24:91–95.

42. Ong J, Sarma J, Venkataraman K, Levin S, Singh B: Circadian rhythmicity of heart rate and QTc interval in diabetic autonomic neuropathy: implications for the mechanisms of sudden death. Am Heart J 1993; 125:744–752.

43. Sarma J, Venkataraman K, Nicod P, et al: Circadian rhythmicity of rate-normalized QT interval in hypothyroidism and its significance for development of class III antiarrhythmic agents. Am J Cardiol 1990; 66:959–963.

44. Schwartz PJ, Wolf S: QT interval prolongation as predictor of sudden death in patients with myocardial infarction. Circulation 1978; 57:1074–1077.

45. Haynes RE, Hallstrom AP, Cobb LA: Repolarization abnormalities in survivors of out-of-hospital ventricular fibrillation. Circulation 1978; 57:654–658.

46. Anhve S, Gilpin E, Madsen EB, Froelicher V, Henning H, Ross J Jr: Prognostic importance of QTc interval at discharge after acute myocardial infarction: a multicenter study of 865 patients. Am Heart J 1984; 108:395–400.

47. Algra A, Tijssen JGP, Roelandt J, Pool J, Lubsen J: QTc prolongation measured by standard 12-lead electrocardiography is an independent risk factor for sudden death due to cardiac arrest. Circulation 1991; 83:1888–1894.

48. Moller M: QT interval in relation to ventricular arrhythmias and sudden

cardiac death in post-myocardial infarction patients. Acta Med Scand 1981; 220:73–77.

49. Wheelam K, Mukharji J, Rude RE: Sudden death and its relation to QT prolongation after acute myocardial infarction: 2 year follow-up. Am J Cardiol 1986; 57:745–750.

50. Boudoulas H, Sohn YH, O'Neill WM, Brown R, Weissler AM: The QT<QS2 syndrome: a new mortality indicator in coronary artery disease. Am J Cardiol 1982; 50:1229–1235.

51. Pohjola-Sintonen S, Siltanen P, Haapakosi J: The QTc interval on the discharge electrocardiogram for predicting survival after acute myocardial infarction. Am J Cardiol 1986; 57:1066–1068.

52. Martí V, Guindo J, Homs E, Viñolas X, Bayés de Luna A: Peaks of QTc lengthening in Holter recordings as a marker of life-threatening arrhythmias in postmyocardial infarction patients. Am Heart J 1992; 124:234–235.

53. Laguna P, Thakor NV, Caminal P, Jané R, Bayés de Luna A, Marti V, Guindo J: New algorithm for QT interval analysis in 24-hour Holter ECG: performance and applications. Med Biol Eng Comput 1990; 28:67–73.

54. Algra A, Tijssen JGP, Roetlandt JRTC, Pool J, Lubsen J: QT interval variables from 24 hour electrocardiography and the two year risk of sudden death. Br Heart J 1993; 70:43–48.

5

Circadian Pattern of Myocardial Ischemia during Daily Life:

Pathophysiological Basis and Therapeutic Considerations

Prakash C. Deedwania, MD

Introduction

Myocardial ischemia is the cardinal manifestation of coronary artery disease (CAD). Although in the past the clinician largely relied on symptoms of angina to assess the clinical impact of CAD, there is now considerable evidence to suggest that anginal symptoms are not a reliable marker of the degree of myocardial ischemia and severity of CAD.[1–3] During the past two decades, a large number of investigations have been performed using ambulatory ECG monitoring (AEM) techniques to evaluate for myocardial ischemia during daily life in patients with CAD.[1–5] These studies have demonstrated that most episodes of myocardial ischemia occur without associated anginal symptoms (now commonly referred to as "silent myocardial ischemia").[1] Several recent

From: Deedwania PC (ed): *Circadian Rhythms of Cardiovascular Disorders.*
©Futura Publishing Co., Inc., Armonk, NY, 1997.

studies have shown that it is the presence of myocardial ischemia, regardless of associated symptoms, that determines the subsequent outcome in patients with CAD.[3–7] The use of Holter monitoring has also
been extremely revealing in the evaluation of chronobiology of myocardial ischemia in patients with CAD.[8–13] A number of studies have
demonstrated that, similar to the circadian rhythm of acute myocardial
infarctions, sudden cardiac death, and cardiac arrhythmias, the episodes
of transient myocardial ischemia also have a circadian pattern.[8–14] This
chapter will focus on the available data supporting chronobiology of myocardial ischemia with special emphasis on the underlying pathophysiological processes that are responsible for the circadian pattern of myocardial ischemia.

Circadian Pattern of Myocardial Ischemia

During the last decade, a number of studies have been performed
to evaluate the prevalence and pattern of myocardial ischemia during
daily life in patients with stable CAD.[8–12] Most of these studies have
identified that ischemic episodes during daily life show a circadian
rhythm with greatest density of ischemia during the period between 6
AM and noon (Fig. 1).[8–14] The circadian pattern of myocardial ischemia
is quite similar to that described (see Chapters 1 and 2) for acute myocardial infarction (MI), sudden death, and cerebrovascular accidents.
The circadian pattern of myocardial ischemia parallels the circadian
rhythms of heart rate, blood pressure, cardiac contractility (the three
major determinants of myocardial oxygen demand), and the sympathetic activity, all of which show significant increases in the morning
hours.[9] Based on the changes in these pathophysiological correlates,
there is a general belief that patients with CAD are at a substantial risk
of an acute coronary event and myocardial ischemia during the morning hours. It is conceivable that once an episode of myocardial ischemia
occurs due to increased myocardial oxygen demand in the morning
hours, it can set up a self-perpetuating cycle because most ischemic
episodes result in further increases in heart rate and blood pressure. It
is also conceivable that further increases in heart rate, blood pressure,
and sympathetic activity can lead to plaque fissuring or rupture of the
vulnerable atherosclerotic plaque, which can create an environment for
acute coronary thrombosis and the resultant MI.[15–20] The increased
platelet aggregability and the decreased intrinsic fibrinolytic activity
during the morning hours further enhance the environment for such an
event.[19,20] Although teleologically the idea of such an ischemic cascade
is appealing, there is currently little evidence to support this concept.

Circadian Variation of Myocardial Ischemia

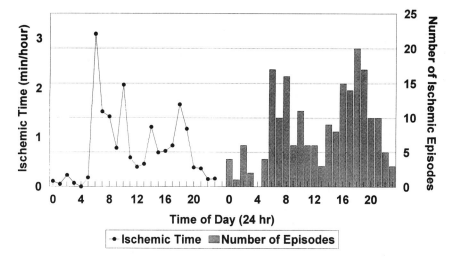

Figure 1: The hour-by-hour circadian variation of myocardial ischemia in 63 CAD patients is depicted here. The total ischemic time (minutes) in each hour of the day is displayed on the left, and the number of ischemic episodes in each hour on the right. (Adapted with permission from reference #14.)

Future studies need to focus further research in these areas to better define the pathophysiological process involved in the onset of acute coronary events.

Based on the data available from the studies utilizing continuous ECG monitoring for 24–48 hours, it is estimated that between 6:00 AM and noon there is a twofold to threefold increase in the risk of myocardial ischemia in patients with CAD (Fig. 1).[9–11] The precise reason for this increase in the ischemic activity in morning hours is not well understood but it is quite likely that the increase in myocardial oxygen demand due to sudden augmentation in sympathetic activity with awakening and resumption of upright posture plays a dominant role. In addition, it is known that other stimuli, such as the effects of the first cigarette in smokers (after an overnight nicotine withdrawal), physical activities, and mental stress in the morning time also contribute to the special predilection for ischemic episodes in the morning hours.[11,13,14]

Most of the previous studies evaluating the circadian rhythmicity of myocardial ischemic have emphasized the clustering of ischemic episodes in the morning hours by showing that as much as 40% of ischemic episodes were observed during the morning hours between 6:00

AM and noon.[8–12] These studies have also noted that the ischemic episodes that occur during the morning hours are generally more severe and last longer when compared to the ischemic episodes that occur spontaneously during other times of the day.[9,11] Recently, the question has been raised as to whether it is a specific time of the day or the period of awakening and the subsequent physical activities that increases the risk of myocardial ischemia in the morning hours. The results of a recent study are quite revealing in this regard.[12] In this study, the investigators monitored a group of patients with stable CAD during a regular activity day (awaken and assume activity at 8:00 AM) and a delayed activity day (awaken at 8:00 AM, arise at 10:00 AM, and begin activity at noon).[12] The results of ambulatory ECG monitoring on the regular activity day demonstrated the morning increase in heart rate and the number of ischemic episodes.[12] In contrast, however, the monitoring during the delayed activity day revealed that corresponding with the delay in resumption to upright posture and physical activities, there was a 4-hour lag phase in the increases in heart rate and increased risk of ischemic episodes. These data are quite intriguing and do indeed suggest that resumption of upright posture and physical activities might be of greater importance than a specific time of the day for the circadian pattern of myocardial ischemia.[12] Because most people are subjected to a considerable amount of physical and mental stress after waking up from sleep and resuming normal activities in the morning, it should not be surprising that the risk of myocardial ischemia is substantially increased during the morning hours. Further work is needed to get insight into specific mechanisms responsible for the circadian pattern of myocardial ischemia.

Pathophysiological Basis for Circadian Pattern of Ischemia

Although this field is still in its evolutionary stage, it has become quite clear that, despite the importance of the time of the day, it is the presence and/or activation of various intrinsic and extrinsic triggers that appears to be primarily responsible for episodes of myocardial ischemia.[19,20] It is also known that a trigger might cause an ischemic event at one time but the same trigger in that given individual at a different time and setting might not result in any ischemic event.[20] Such variable influences of triggers in different settings and during various times of the day suggest that the precise mechanism by which a specific trigger initiates an ischemic event might be quite complex and an interplay between a variety of factors other than the trigger might be as important as the presence or activation of a given trigger. Based on the available data, it is evident that a clear understanding of the pathophysiological processes responsible for ischemic events and the specific role played by various triggers would be helpful in better defining the therapeutic strategies for prevention and treatment of ischemia episodes in the morning as well as other times of the day.

Hemodynamic Changes and Circadian Pattern

It is well recognized that an episode of myocardial ischemia is usually triggered whenever there is a significant increase in heart rate and/or blood pressure which exceeds the ischemic threshold for a given patient.[9,21,22] It is, however, important to recognize that the ischemic threshold might vary from one time to another and is significantly influenced by the level of coronary vasomotor tone.[23,24] In the early morning hours when there is an increased vasomotor tone, the ischemic threshold, measured by using heart rate as a surrogate, is usually lower when compared to the ischemic threshold in midafternoon when vasomotor tone is reduced.[25] This example clearly suggests that although the changes in hemodynamic determinants of myocardial oxygen demand have an important role, it is the interplay between various factors that plays a significant role in the pathophysiological process that finally determines whether an ischemic event is going to occur or not.[26,27]

It is known that patients with coronary artery disease are exposed to various stimuli and stressful situations that could increase heart rate and blood pressure levels on multiple occasions throughout a given day, but ischemic episodes do not occur in every one of these patients on each of these occasions. This emphasizes that there must be an interplay between a constellation of factors for an event to occur. Herein lies the importance of the time of day and activation of various factors. It is known that the morning time is particularly important because, in addition to the surge in heart rate, blood pressure and cardiac contractility, the vasomotor tone is enhanced and the platelet aggregability is also increased during this period.[9,28]

Despite the role of various other factors in the pathogenesis of myocardial ischemia, it should be noted that changes in hemodynamic parameters are of paramount importance in initiating most ischemic episodes. There are a number of factors that can influence the changes in hemodynamic factors involved in the pathogenesis of myocardial ischemia. These include time of day, sleep/wake cycle, changes in posture, physical activities including vigorous exercise, mental stress and anger, smoking, and periods of rapid eye movement (REM) sleep.[9–13,29–34] The precise mechanism by which hemodynamic changes occur in various settings as well as the exact nature and degree of hemodynamic changes observed might vary considerably during a specific time of day. It is well known that the risk of myocardial ischemia is significantly increased during the morning hours between 6 AM and noon. Previous studies had indicated that the increased risk of ischemic episodes in the morning hours is related to the biological rhythms based on time of day and the associated changes in various intrinsic hormonal and hemostatic factors as well as the sympathetic activity.[19,20] More recently, however, the results of several studies have demonstrated that it is not a specific time of the day but the associated changes in posture and the activity level that determine the changes in hemodynamic and hemostatic factors responsible for ischemic events.[13,14,20] Because most people sleep during the night time and wake up in the morning, a variety of changes that occur upon awakening might indeed be responsible for the well-demonstrated increase in myocardial ischemia in the morning. It would be useful to examine the specific role of the most important factors that contribute to the circadian pattern of myocardial ischemia.

Increased Sympathetic Activity and Risk of Ischemia in the Morning

During sleep, sympathetic activity is usually at its minimum and heart rate and blood pressure are also significantly reduced.[31] However, upon awakening, sympathetic activity increases substantially and it is associated with a significant surge in heart rate and blood pressure.[9,20] Because most people with a normal sleep pattern wake up between 5 AM and 8 AM, the associated surge in the sympathetic activity can result in significant changes in the hemodynamic parameters (e.g., heart rate, blood pressure, vasomotor tone, etc.) in the morning, resulting in increased risk of myocardial ischemia during this time period. The significance of the hemodynamic changes in the morning hours has been evaluated by 24-to-48-hour simultaneous ECG and blood pressure monitoring in patients with coronary artery disease, while patients continued their normal daily routine (including sleep/ wake cycles) and maintained their usual activities.[8,9] In these studies, it was observed that the number of myocardial ischemic episodes abruptly increased in the morning hours and between 30% and 40% of the total ischemic activity occurred between the hours of 6 AM and noon.[9–12] The evaluation of heart rate, systolic blood pressure, and the calculated double product (systolic BP X HR) showed simultaneous and abrupt increases that paralleled the increased ischemic activity (Fig. 2), clearly establishing the role of hemodynamic changes in the pathogenesis of the increased risk of ischemic episodes during the morning hours.[9] Whether the observed changes occurred primarily as a result of

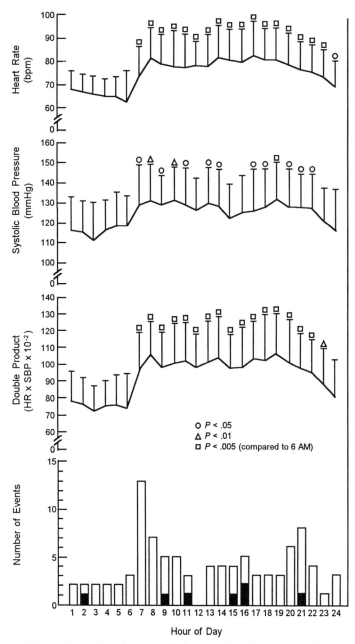

Figure 2: The 24-hour distribution of silent ischemic events during continuous electrocardiographic monitoring parallels the circadian pattern of heart rate (HR) and systolic blood pressure (SBP). Note the simultaneous and significant increase in both the circulatory changes and silent ischemic events between 0600 h and 0700 h. (Adapted with permission from reference #9.)

changes governed by an intrinsic biological clock or were caused by sleep/wake cycle and the postural changes could not be determined from the results of these studies. The results of a recent study, however, demonstrated that during two separate monitoring sessions when patients were randomly asked to delay the onset of physical activity during one of the sessions, the peak in the risk of ischemic episodes was also delayed and correlated with the time of activity.[12] An earlier study had also suggested that the increases in catecholamine levels and platelet aggregability were more precisely related to the resumption of upright posture (after waking up) rather than to the time of awakening.[28] The precise reasons for the increased sympathetic activity and the related changes in hemodynamic parameters associated with resumption of upright posture have not been fully established. It has been postulated that resumption of upright posture (after a prolonged period of sleep/rest) and gravitational forces lead to blood pooling in the venous beds in lower extremities, which results in decreased venous return to the heart with subsequent decrease in left ventricular volume and preload resulting in reduced stroke volume. The resultant fall in arterial blood pressure then causes an increase in heart rate and cardiac contractility by activating the sympathetic nervous system in an attempt to restore hemodynamic stability. The resumption of physical activities might also play a role. It is well known that changes in plasma catecholamines are directly related to the level of physical activity. These data clearly emphasize the fact that it is not just the specific time of day but the related changes in posture and activity level that are also of crucial importance in determining the risk of myocardial ischemia at any given time of the day.

Physical Activities as Triggers of Ischemia

The importance of physical activity as a trigger of transient myocardial ischemia has also been recently emphasized.[13] In a study of patients with proven CAD who were monitored by continuous Holter monitoring during two separate recording sessions, it was demonstrated that the peak ischemic activity correlated more closely with the time of physical activity than a specific time of the day.[13] In some studies it was also demonstrated that treatment with the beta blockers effectively suppressed the increased risk of myocardial ischemia secondary to physical activity.[10,13] These data suggest that the increase in myocardial oxygen demand during physical activity plays a significant role in the genesis of myocardial ischemia.[9,14,22] However, it should be noted that reduction in coronary blood supply due to vasoconstriction of atherosclerotic coronary arteries in response to dynamic exercise might also occur and could also play a role in the pathogenesis of myocardial ischemia induced by physical activity.[23–25]

Mental Stress as a Trigger of Myocardial Ischemia

Although there is only a limited amount of information available from well-designed, controlled studies, anecdotally everyone has heard of, or encountered, cases of myocardial infarction, sudden cardiac death, and episodes of angina experienced during periods of mental stress and anger.[14,25] It is known that during periods of mental stress there is not only an increase in heart rate and blood pressure, but coronary blood flow might also be significantly reduced due to enhanced alpha-adrenergic activity and the associated increase in vasomotor tone.[14,32,33] In an experimental model, Verrier and coworkers demonstrated that coronary vascular resistance was significantly increased and resulted in decreased coronary blood flow during episodes of anger.[32] Thus, it is conceivable that mental stress can precipitate myocardial ischemia by a variety of mechanisms that may be activated concomitantly. A number of investigators have evaluated the mechanism by which mental stress induces myocardial ischemia.[14,29,32,33] It is well known that mental stress is one of the most potent stimuli that can affect hemodynamic variables in much the same way as physical stress and can result in increased myocardial oxygen demand that can eventually lead to myocardial ischemia.[14,29,34] Several controlled observations have shown that, in patients with coronary artery disease, mental stress frequently induces myocardial ischemia.[14,29,34] Some of these studies have performed careful evaluations of simultaneous electrocardiographic and blood pressure monitorings and demonstrated a significant increase in both heart rate and blood pressure during periods of mental stress, although the magnitude of increase in these parameters differed between various studies.[34,35] In general, the mental stressors (e.g., tasks involving speech and mathematic challenge) produced significant increases in heart rate and blood pressure, although the magnitude of changes in heart rate was considerably less than that observed during exercise testing.[34] However, when compared with changes during exercise testing, the increase in diastolic blood pressure was of much greater magnitude during mental stress whereas the increase in systolic blood pressure was quite comparable in the two settings.[34,35] The calculated double product was generally lower during mental stress-induced ischemia when compared with the double product during exercise induced ischemia.[34,35] These comparisons of hemodynamic changes do indeed suggest that the mental stress-induced ischemia might be produced by a combination of factors that includes both increased myocardial oxygen demand as well as reduced coronary blood flow secondary to coronary vasospasm.[14,34,35] However, the significant increases in both systolic and diastolic blood pressure during mental stress clearly support the dominant role of increased myocardial oxygen demand in the genesis of myocardial ischemia induced by mental stress.[34,35]

Interactions between Physical Activities, Mental Stress and Endogenous Factors

It is well known that at times there are periods during which both the level of physical activities and the degree of mental stress are increased together, and such periods might indeed predispose a patient with CAD to increased risk of myocardial ischemia. However, it is also known that patients do not always experience ischemic signs and/or symptoms during all such encounters. Thus, it is conceivable that the endogenous circadian component might indeed play a role in increasing the risk of ischemia during a specific period of time in conjunction with increased physical activity or mental stress.[14] A recent study in a large group of patients with well-defined CAD has carefully evaluated the interrelationships between these factors and provides important insight into the pathophysiological processes involved in the circadian pattern of myocardial ischemia.[14] The results of this study revealed that a close relationship exists between the morning increase in myocardial ischemic episodes and the increase in physical activities and mental stress encountered during the corresponding period (Fig. 3).[14] It was shown that the increased susceptibility for myocardial ischemia was present for 2 hours after the resumption of upright posture upon awakening.[14] The investigators also correlated the risk of morning ischemia with the changes in heart rate and demonstrated that the increased risk of ischemia in morning hours persisted even after controlling for the mean hourly heart rate and heart rate increases before the onset of ischemic episodes.[14] Based on these data, the investigators have suggested that, in addition to the obvious importance of increased myocardial oxygen demand secondary to physical activities and mental stress, endogenous factors such as enhanced vasomotor tone and/or changes in hemostatic factors might also play an important role in the chronobiology of myocardial ischemia.[14] Clearly more work is needed in this area to better define the role of various pathophysiological factors in the circadian rhythmicity of myocardial ischemia.

Diurnal Variation of Physical and Mental Activity as Related to Ischemia

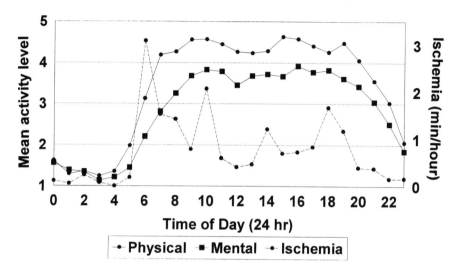

Figure 3: Diurnal variation of physical and mental activities as related to total ischemic time per hour, displayed in 1-hour intervals throughout the day. The significant morning increase in ischemic time was paralleled by an elevation of physical and mental activity levels, but the subsequent decrease in ischemic time around noon was not accompanied by a decrease in mean level of activity. In the late evening, activity levels and ischemia decreased to their lowest levels. (Adapted with permission from reference #14.)

Therapeutic Considerations

Because the last two chapters in this book provide specific details regarding therapeutic considerations for circadian periodicity of cardiovascular events, this section will include only a brief discussion pertinent to therapeutic approaches in myocardial ischemia. Based on the preceding discussion regarding the pathophysiological role of various factors in the circadian periodicity of myocardial ischemia, it should be evident that certain therapies would be more effective than others in reducing the risk of ischemia in the morning hours.[10,11] It should, however, be noted that, although some treatments (e.g., beta blockers) may be better for suppressing ischemia in the morning hours, other treatment (e.g., calcium antagonists) might be more effective for ischemia during the evening and night time.[10–13,36–39] That this is indeed the case was recently emphasized by detailed analyses of heart rate and ischemic episodes in a double-blind, crossover trial that indicated differential effects of various therapeutic agents on ischemic episodes during various times of the day.[21] The results of this study revealed that, in general, most episodes during the day are preceded by increases in heart rate and therefore treatment with beta-blocking agents is most effective for the vast majority of ischemic episodes. However, for the minority of ischemic episodes that are not preceded by increases in heart rate (most of which occur during night time), it was demonstrated that treatment with a calcium channel blocker might be preferable because coronary vasoconstriction might be playing a dominant role in the pathogenesis of the low heart rate episodes.[21,26,27] Clearly, then, it is difficult to suggest that a single agent can suppress all episodes of myocardial ischemia throughout the day. Because a combination of factors might play a role in the pathogenesis of ischemic episodes during various times of the day, it might be necessary to use combination therapy with drugs that reduce myocardial oxygen demand as well as those that improve the coronary blood flow by the vasodilating actions on coronary vasculature.[27] The precise selection of initial therapy would vary, depending upon the dominant mechanism playing the role in pathogenesis of ischemia in a given patient.

A number of studies have evaluated the efficacy of various therapies on circadian rhythmicity of myocardial ischemia.[10,11,13,36–39] The results of these studies have demonstrated that, although most anti-ischemic drugs are effective in reducing the frequency and duration of ischemia, treatment with beta blockers is clearly most effective and superior to other drugs in suppressing the ischemic episodes in the morning hours.[10,11,13,36–38] The superior efficacy of beta blockers in reducing the risk of morning ischemia is consistent with the dominant pathophysiological role of increased sympathetic activity and the associated increases in heart rate and blood pressure during the morning hours.[9,26,27] The discussions in Chapter 11 provide further insight into the mechanism of action of beta blockers; however, it is important to emphasize here that the reduction in heart rate and prevention of the morning surges in heart rate and blood pressure by beta blockers are indeed the main mechanisms responsible for their superior efficacy in abolishing the ischemia.[9,36–38] Although more work is needed, the concept of chronotherapeutics is quite intriguing (see Chapter 10) and suggests that the long-acting beta blockers given at bedtime might be the most prudent therapeutic approach to minimize the increased risk of ischemia in the morning hours. It has been suggested that such a therapeutic strategy might also be beneficial in reducing the risk of acute MI and sudden death in the morning hours.

In addition to the beneficial effects of beta-blocking agents, there is evidence to suggest that treatment with nitrates and calcium channel blockers might also be of therapeutic benefit, especially in suppressing ischemic episodes that are not preceded by increases in heart rate or blood pressure.[11,21,22,39] Some studies have shown that, when used in combination with beta blockers, the use of calcium antagonists is associated with further reduction in the frequency and duration of ischemic episodes.[39] In this regard, the recent introduction of time-release preparations of verapamil (Covera HS), which is administered at bedtime and releases the drug slowly in the first few hours but provides therapeutic levels of the drug before the time of awakening, is quite intriguing and provides a rational therapeutic basis for delivering the drug when it is most needed.[40,41] If the ongoing clinical trials prove the therapeutic superiority of such a chronotherapeutic approach, it would be reasonable to suggest that future treatment of disorders that demonstrate circadian periodicity would be improved considerably with the hope of reducing the risk of coronary events and cardiac death.

References

1. Deedwania PC, Carbajal EV: Silent myocardial ischemia: a clinical perspective. Arch Intern Med 1991; 151:2373–2382.
2. Cohn PF: Silent myocardial ischemia. Ann Int Med 1988:109; 312–317.
3. Gottlieb SO, Weisfeldt ML, Ouyang P, Mellits ED, Gerstenblith G: Silent ischemia as a marker for early unfavorable outcomes in patients with unstable angina. N Engl J Med 1986; 314:1214–1219.
4. Deedwania P, Carbajal E: Prevalence and patterns of silent myocardial ischemia during daily life in stable angina patients receiving conventional antianginal drug therapy. Am J Cardiol 1990; 65:1090–1096.
5. Pepine CJ, Geller NL, Knatterud GL, et al: The Asymptomatic Cardiac Ischemia Pilot (ACIP): design of a randomized clinical trial, baseline data and implications for a long-term outcome trial. J Am Coll Cardiol 1994; 24:1–10.
6. Deedwania P, Carbajal EV: Silent ischemia during daily life is an independent predictor of mortality in stable angina. Circulation 1990; 81: 748–756.
7. Deedwania PC: Asymptomatic ischemia during pre-discharge Holter monitoring predicts poor prognosis in the postinfarction period. Am J Cardiol 1993; 71:859–861.
8. Rocco MB, Barry J, Campbell S, et al: Circadian variation of transient myocardial ischemia in patients with coronary artery disease. Circulation 1987; 75:395–400.
9. Deedwania PC, Nelson J: Pathophysiology of silent myocardial ischemia during daily life. Circulation 1990; 82:1296–1304.
10. Deedwania PC, Carbajal EV, Nelson JR, et al: Anti-ischemic effects of atenolol versus nifedipine in patients with coronary artery disease and ambulatory silent ischemia. J Am Coll Cardiol 1991; 17:963–969.

11. Pepine CJ: Circadian variations in myocardial ischemia: implications for management. JAMA 1991; 265(3):386–390.
12. Deedwania PC, Pepine CJ, Cohn P, et al, for the ASIST Study Group: Prospective evaluation of circadian pattern of transient myocardial ischemia in asymptomatic and minimally symptomatic patients with coronary artery disease. E Heart J 1993; 14:136.
13. Parker J, Testa M, Jimenez A, et al: Morning increase in ambulatory ischemia in patients with stable coronary artery disease: importance of physical activity and increased cardiac demand. Circulation 1994; 89:604–614.
14. Krantz DS, Kop WJ, Gabbay FH, et al: Circadian variation of ambulatory myocardial ischemia. Circulation 1996; 93:1364–1371.
15. Fuster V, Badimon L, Badimon JJ, et al: The pathogenesis of coronary artery disease and the acute coronary syndromes (2). N Engl J Med 1992; 326:310–318.
16. Gertz SD, Roberts WC: Hemodynamic shear force in rupture of coronary arterial atherosclerotic plaques (editorial). Am J Cardiol 1990; 66:1368–1372.
17. Muller JE, Stone PH, Turi ZG, et al: Circadian variation in the frequency of onset of acute myocardial infarction. N Engl J Med 1985; 313:1315–1322.
18. Muller JE, Ludmer PL, Willich SN, et al: Circadian variation in the frequency of sudden cardiac death. Circulation 1987; 75:131–138.
19. Muller JE, Tofler GH: Circadian variation and cardiovascular disease. N Engl J Med 1991; 325:1038–1039.
20. Muller JE, Abela GS, Nesto RW, et al: Triggers, acute risk factors and vulnerable plaques: the lexicon of a new frontier. J Am Coll Cardiol 1994; 23:809–813.
21. Andrews TC, Fenton T, Toyosaki N, et al, for the Angina and Silent Ischemia Study Group (ASIS). Subsets of ambulatory myocardial ischemia based on heart rate activity: circadian distribution and response to anti-ischemic medication. Circulation 1993; 88:92–100.
22. Benhorin J, Shmuel B, Moriel M, et al: Circadian variations in ischemic threshold and their relation to the occurrence of ischemic episodes. Circulation 1993; 87:808–814.
23. Panza JA, Epstein SE, Quyyumi A, et al: Circadian variation in vascular tone and its relation to alpha sympathetic vasoconstrictor activity. N Engl J Med 1991; 325:986.
24. Gaze JE, Hess OM, Murakami T, et al: Vasoconstriction of stenotic coronary arteries during dynamic exercise in patients with classic angina pectoris: reversibility by nitroglycerine. Circulation 1986; 73:865.
25. Quyyumi AA, Panza JA, Diodata JG, et al: Circadian variation in ischemic threshold: a mechanism underlying the circadian variation in ischemic events. Circulation 1992; 86:22–28.
26. Deedwania PC: Increased myocardial oxygen demand and ischemia during daily life: resurrection of an age-old concept. J Am Coll Cardiol 1992; 20(5):1099–1100.
27. Deedwania PC: Increased demand versus reduced supply and the circadian variations in ambulatory myocardial ischemia: therapeutic implications. Circulation 1993; 88:318–331.
28. Brezinski DA, Tofler GH, Muller JE, et al: Morning increase in platelet aggregability: association with assumption of upright posture. Circulation 1988; 78:35–40.

29. Barry J, Selwyn AP, Nabel EG: Frequency of ST-segment depression produced by mental stress in stable angina pectoris from coronary artery disease. Am J Cardiol 1988; 61:989–993.

30. Belch JJ, McArdle BM, Burns P, et al: The effects of acute smoking on platelet behaviour, fibrinolysis and haemorheology in habitual smokers. Thromb Haemost 1984; 51:6–8.

31. Somers VK, Dyken ME, Mark AL, et al: Sympathetic-nerve activity during sleep in normal subjects. N Engl J Med 1993; 328:303–307.

32. Verrier RL, Hagestad EL, Lown B: Delayed myocardial ischemia induced by anger. Circulation 1987; 75:249–254.

33. Yeung AC, Vekshtein VI, Krantz DS, et al: The effect of atherosclerosis on the vasomotor response of coronary arteries to mental stress. N Engl J Med 1991; 325:1551–1556.

34. Rozanski A, Bairey C, Krantz D, et al: Mental stress and the induction of myocardial ischemia in patients with coronary artery disease. N Engl J Med 1988; 318:1005–1012.

35. Krantz DS, Helmers KR, Bairey CN, et al: Cardiovascular reactivity and mental stress-induced myocardial ischemia in patients with coronary artery disease. Psychosomatic Med 1991; 53:1–12.

36. Deedwania P, Pepine CJ, Cohn P, et al, for the ASIST Study Group: The morning increase in myocardial ischemia is effectively suppressed by atenolol. Circulation 1993; 88(4 Part 2):1594.

37. Willich SN, Pohjola-Sintonen S, Bhatia SJS, et al: Suppression of silent ischemia by metoprolol without alteration of morning increase of platelet aggregability in patients with stable coronary artery disease. Circulation 1989; 79:557–565.

38. Lambert CR, Coy K, Imperi G, et al: Influence of beta-adrenergic blockade defined by time series analysis on circadian variation of heart rate and ambulatory myocardial ischemia. Am J Cardiol 1989; 64:835–839.

39. Parmley WW, Nesto RW, Singh BN, et al: Attenuation of the circadian patterns of myocardial ischemia with nifedipine GITS in patients with chronic stable angina. J Am Coll Cardiol 1992; 19:1380–1389.

40. White WB, Anders RJ, MacIntyre JM, et al, and the Verapamil Study Group. Nocturnal dosing of a novel delivery system of verapamil for systemic hypertension. Am J Cardiol 1995; 76:375–380.

41. Cutler NR, Anders RJ, Jhee SS, et al: Placebo controlled evaluation of three doses of a controlled-onset, extended-release formulation of verapamil in the treatment of stable angina pectoris. Am J Cardiol 1995; 75:1102–1106.

6

Circadian Variation in the Onset of Myocardial Infarction

James E. Muller, MD

Introduction

It is now well established that onset of myocardial infarction (MI) has a distinct daily pattern with a peak incidence in the hours after awakening and arising.[1] A surprising feature of this morning peak in onset of MI is that it was not widely appreciated or accepted as recently as 1985. Prior to that time there had been several reports of a morning increase of MI onset,[2,3] including a large WHO study,[4] but these findings were not widely accepted. The primary reason for lack of acceptance appears to have been the reliance in these studies on symptoms of infarction to recognize the time of onset. Skeptics could argue that a morning increase in the reporting of the pain of infarction would be expected even with a random distribution of onset over 24 hours—patients might sleep through the onset of infarction and recognize the symptoms only after awakening.

Documentation that the onset of MI demonstrates a prominent circadian rhythm required the use of a method to eliminate a "morning discovery" explanation for the findings. The method used was reliance on creatine kinase timing to determine objectively the time of onset of infarction.[5,6] Two studies used this method to overcome the objections

From: Deedwania PC (ed): *Circadian Rhythms of Cardiovascular Disorders.*
©Futura Publishing Co., Inc., Armonk, NY, 1997.

stated above to the reports[2-4] that used onset of pain as the marker for time of MI onset. In the Multicenter Investigation of Limitation of Infarct Size (MILIS),[5] numerous creatine kinase MB determinations were made to obtain an objective determination of the time of onset of MI in 849 patients. The onset of MI was considered to have occurred 4 hours before the initial elevation of creatine kinase. The time of day of MI onset in the 703 patients for whom complete creatine kinase timing was available is shown in Figure 1. The abscissa shows the hour of the day with the same data replotted for a second day to permit appreciation of the relation between the end of one day and the beginning of the next. The ordinate shows the number of infarcts per hour. A marked daily variation is present, with a maximum of 45 infarcts between 9 AM and 10 AM and a minimum of 15 between 11 PM and midnight. This objective evidence obtained in the MILIS database was confirmed in the database (4,741 patients) from the Intravenous Streptokinase in Acute Myocardial Infarction Study[6] demonstrating that MI was four times more likely to occur between 8 AM and 9 AM than between midnight and 1 AM (Fig. 2).

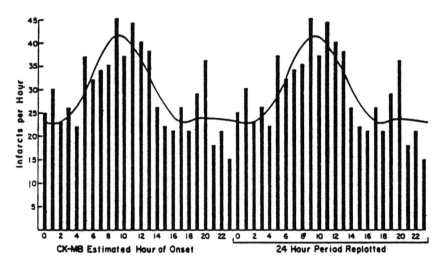

Figure 1: Hourly frequency of onset of myocardial infarction as determined by the creatine kinase-MB method. The number of infarctions beginning during each of the 24 hours of the day is plotted on the left side of the figure. On the right, the identical data are plotted again to permit appreciation of the relation between the end and the beginning of the day. A two-harmonic regression equation for the frequency of onset of myocardial infarction has been fitted to the data (curved line). A primary peak incidence of infarction occurs at 9 AM and a secondary peak occurs at 8 PM. (Reprinted with permission.[1])

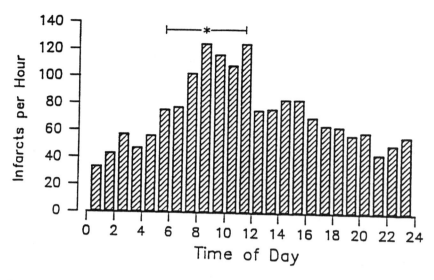

Figure 2: Bar graph of incidence of myocardial infarction of 1,741 patients of the ISAM (Intravenous Streptokinase in Acute Myocardial Infarction) Study. There is a marked circadian variation (p <0.001) with a peak during the morning hours. Myocardial infarction occurred 1.8 times more frequently between 6 AM and 12 noon compared with the average of other quarters of the day. The risk of myocardial infarction in the afternoon and evening was approximately equally distributed, whereas during the night, a trough period occurred in the incidence of myocardial infarction. (Reprinted with permission.[6])

The finding of these two studies is further supported by Goldberg et al.[7] and Willich et al.[8] who have subsequently refined this evidence by determining that the increased incidence of MI occurs within the first 3 to 4 hours after awakening and onset of activity.

All of the preceding studies have been based primarily on the time of onset of Q-wave infarction. Non-Q-wave infarction has been reported to lack a morning increase of occurrence.[9] The finding that PAI-1 levels are lower in patients with non-Q-wave MI than in those with Q-wave MI[10] may indicate the marked daily fluctuation in PAI-1 has less of an influence on non-Q-wave MI than on Q-wave MI.

The findings that numerous other cardiovascular diseases, including transient myocardial ischemia and stroke, also show a morning increase reinforces the conclusion that nonfatal MI has a prominent morning increase in onset. A smaller, evening peak of MI appears in some databases; its cause is not known and requires further investigation.

Autopsy and Angiographic Data Pertinent to Causes of the Morning Increase of Infarction

In 1980, DeWood et al. demonstrated that occlusive coronary artery thrombosis is the cause of most Q-wave MIs.[11] Furthermore, coronary angiographic and angioscopic studies have demonstrated a high frequency of nonocclusive coronary thrombosis in patients with unstable angina.[12,13] Constantinides observed that thrombus had formed over a ruptured atherosclerotic plaque in all of the cases of occluded coronary arteries that he examined.[14] The angiographic finding that contrast media outpouching, indicative of plaque rupture, was observed in patients who had undergone successful thrombolysis supports the importance of plaque rupture in acute coronary syndromes.[15]

Although autopsy studies generally reveal severe atherosclerotic stenosis at the base of a fatal coronary thrombus,[16] there is angiographic evidence that in many patients surviving an MI, the degree of stenosis, as assessed by angiography, is relatively mild and that obstructive thrombus accounts for the majority of the obstruction to blood flow.[17–19] These findings may explain the absence of prior symptoms in many patients presenting with acute MI.

Intrinsic plaque characteristics and extrinsic factors that predispose and initiate plaque disruption remain areas of intense investigation. Richardson and coworkers have reported that in 63% of the cases, plaque disruption occurred at the junction of a lipid pool with normal tissue.[20] Presence of a lipid core, a thin fibrous cap, and macrophage activity seem to be important factors that predispose an atherosclerotic plaque to disrupt.[21] The recent finding that there is a spectrum of lesions from fibrous plaques composed predominantly of smooth muscle cells to lipid-rich lesions with numerous macrophages leads to the concept that inflammatory mechanisms modulate plaque morphology.[22–24]

Morning Increase of Physiological Processes that Might Trigger Myocardial Infarction

A variety of mechanisms, alone or in combination, could account for the morning increase in MI onset. A morning arterial pressure surge could initiate plaque disruption. An increase in coronary arterial tone could worsen the flow reduction produced by a fixed stenosis. The combination of increases in arterial pressure and increases in coronary tone could elevate shear stress (force directed against the endothelium resulting from increased coronary blood flow velocity), thus disrupting a vulnerable plaque. Other prothrombotic processes including increased platelet adhesion, increased blood viscosity, and increased platelet aggregability[25] have been implicated in the morning increase in ischemic events. Such a thrombotic tendency added to a reduced fibrinolytic activity in the morning[26] could increase the likelihood that an otherwise harmless mural thrombus overlying a small plaque fissure would propagate and occlude the coronary lumen.

Although a 24-hour periodicity of disease onset (Figs. 1, 2) and physiological processes is well established, it remains unclear if this periodicity results from a true endogenous circadian rhythm or from the daily rest-activity cycle.[27] Cortisol secretion, for example, is well known to be an endogenous circadian process not dependent on daily activity,[28] while the morning platelet aggregability increase is abolished if the subjects remain at bedrest.[29] There may also be an interaction between circadian and rest-activity cycles, e.g., assumption of the upright posture leading to sympathetic activation may be more likely to cause intense vasoconstriction when endogenously controlled cortisol levels are high. This concept is unresolved because circadian stage and sleep-wake cycles with posture change are highly correlated in all epidemiological studies of infarction timing thus far reported. Investigations analyzing the relationship between an unusual wake time on the day of MI and standard wake time on other days could potentially separate the wake-sleep cycle from the underlying cortisol rhythm, but such studies have not been reported.

Although the peak incidence of disease onset occurs in the morning, it is likely that similar physiological processes trigger disease onset at other times of the day. The peak morning incidence of infarct onset probably results from the synchronization of potential triggers in the morning, while a secondary evening peak in infarct onset observed in the MILIS data[5] and the CAST trial[30] may result from synchronization of the population for an additional potential trigger, such as the evening meal. For other periods of the day, exposure of the population to potential triggers is sporadic, and no other prominent peaks of incidence are observed.

The possibility has been raised that the increased sympathetic activity that accompanies REM sleep might trigger cardiovascular events.[31] This would not account for the prominent morning increase in MI because the peak of the disorder has been shown to increase after awakening. REM-induced sympathetic discharge could, however, account for some of the relatively small number of infarcts that awaken the patient from sleep.

Effect of Drug Therapy on Circadian Variation of Disease

As the field of study of circadian variation of cardiovascular disease has progressed, investigators have attempted to determine if various types of drug therapy alter the timing of cardiovascular events. For nonfatal MI and sudden cardiac death, the conditions that have received the most attention, there is strong evidence that beta-adrenergic blockade, a therapy well-documented to prevent the occurrence of these disorders, selectively decreases the morning peak of events.

The evidence supporting this effect of beta blockade is of two types. First, studies determining the timing of infarction have shown a flattening of the morning peak in patients who happened to be receiving beta-adrenergic blocking agents prior to their infarct.[5,32] Since beta blockers were not randomly assigned, these studies are open to the criticism that the absence of the morning peak is due to confounding by factors other than the therapy. The second type of evidence, which is not subject to the concerns over confounding, comes from the Beta Blocker Heart Attack Trial (BHAT) in which patients were randomly assigned to beta blockade or placebo. Beta blockade demonstrated a selective beneficial effect against the occurrence of sudden cardiac death in the morning.[33]

Evidence indicating a selective benefit of aspirin therapy in the morning has been less impressive than that for beta blockade. Observations of the timing of infarction in patients taking aspirin therapy, but not by random assignment, prior to their infarction have yielded mixed results. However, the single randomized study in which the effect has been studied has demonstrated a selective morning decrease in nonfatal MI in patients receiving aspirin therapy.[34] It is possible that the randomized study is powerful enough to detect a small beneficial effect that cannot be detected by the nonrandomized, observational studies.

Studies of silent myocardial ischemia have demonstrated that beta blockade, but not a short-acting calcium channel blocker, attenuates the morning increase in silent myocardial ischemia.[35] A recent study has demonstrated that the morning increase in silent ischemia can be prevented by nadolol therapy[36] (Fig. 3).

The studies cited above indicate that pharmacological therapy can affect the time of day of cardiac events. An effect of time of day on efficacy of thrombolytic therapy for MI has also been reported.[37,38] Fujita et al. observed a 33% success rate for thrombolysis from 6 AM to noon versus 67% for other times of the day. The authors speculated that the lower success rate in the morning might be due to the formation of more refractory thrombi at that time of day. It has also been reported that infarct size is larger in patients presenting with symptoms during the 6 AM to noon interval.[39]

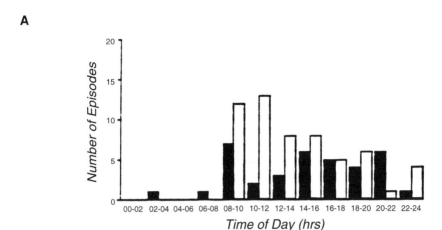

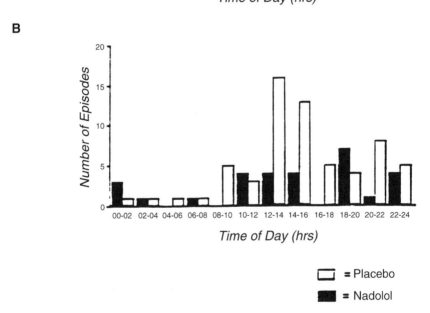

Figure 3: Bar graphs show frequency of episodes of ambulatory ischemia during therapy with placebo and nadolol on the two activity days. **(A)** regular activity day; **(B)** delayed activity day. (Reprinted with permission.[36])

General Theory of Triggering of Coronary Thrombosis

The new information on circadian variation and triggering has provided the basis for a general theory of onset of coronary thrombosis.[40] The hypothesis presented in Figure 4 adds the concept of triggering activities

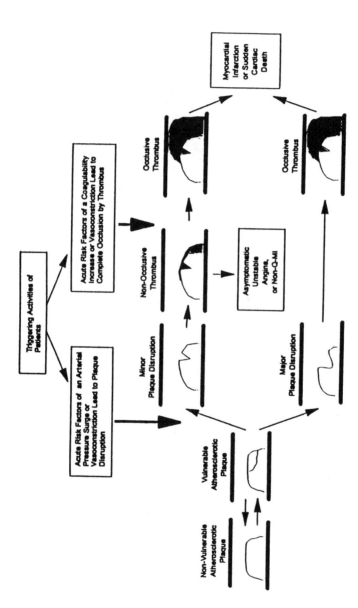

Figure 4: Illustration of a hypothetical method by which daily activities may trigger coronary thrombosis. Three triggering mechanisms are presented: (1) physical or mental stress producing hemodynamic changes leading to plaque rupture, (2) activities causing a coagulability increase, and (3) stimuli leading to vasoconstriction. The scheme depicting the role of coronary thrombosis in unstable angina, myocardial infarction, and sudden cardiac death has been well described by numerous authors. The novel portion of this figure is the addition of triggers. See text for detailed discussion. (Reprinted with permission.[1])

to the general scheme of the role of thrombosis in the acute coronary syndromes advanced by Falk, Davies and Thomas, Fuster, Willerson, and others.[16,41-45] It involves three important new concepts: *triggers, acute risk factors,* and *vulnerable plaques.*

It is postulated that onset occurs when a "vulnerable" atherosclerotic plaque disrupts and occlusive thrombus formation occurs. Hemodynamic stresses may cause the disruption of the plaque; hemostatic and vasoconstrictive forces may then determine if the resultant thrombus is occlusive.

It is proposed that the initial step in the process is the development, with advancing age, of a vulnerable atherosclerotic plaque. Plaque vulnerability is defined functionally as the susceptibility of a plaque to disruption. Development of such vulnerability is a poorly understood process, but is presumably a dynamic, potentially reversible disorder caused by several factors including changes in plaque constituents or its blood supply via vasa vasorum, and/or changes in the functional integrity of the overlying endothelium due in part to increased macrophage activity and thinning of the plaque collagen cap. The new catheterization laboratory techniques of intracoronary angioscopy and ultrasound may, in the future, permit detection of vulnerable plaques prior to their disruption.

Onset of MI might begin when a physical or mental stress produces a hemodynamic change that is sufficient to disrupt a vulnerable plaque. Vasoconstrictive and thrombogenic forces might then lead to coronary occlusion. Shear forces may also play an important role in thrombus formation as suggested by Folts[44] and others.

The finding that disrupted plaques without thrombi are sometimes observed at autopsy in patients dying of noncardiac disease suggests that, in some cases, acute plaque disruption may not be the initial step in disease onset.[46] In such patients, the trigger may lead to occlusive thrombosis by causing an increase in thrombotic tendency, or vasoconstriction, in the presence of a previously nonthrombogenic plaque. It is also possible that the plaques became disrupted at the time of death, an occurrence perhaps more likely during a violent death. These possibilities underscore the need for studies of the lesions causing disease onset in living patients.

A synergistic combination of triggering activities may account for thrombosis in a setting in which each activity alone may not exceed the threshold for causation of infarction. For example, the combination of physical exertion (producing a minor plaque disruption) followed by cigarette smoking (producing an increase in coronary artery vasoconstriction and a relatively hypercoagulable state)[47] may be needed to cause occlusive thrombosis and disease onset. Also, the response to a potential trigger of a healthy individual may differ from that observed in an individual with a condition predisposing to MI. Exaggerated or paradoxical responses may be observed. For example, hypertensives demonstrate a greater increase in forearm vascular resistance after infusion of norepinephrine than normals.[48] Patients with atherosclerosis may demonstrate a paradoxical vasoconstrictor response in response to acetylcholine infusion[49] and an impaired increase in fibrinolytic potential with exercise.[50]

The findings of circadian variation and triggering have also led to the concept of an acute risk factor that supplements the traditional concept of a chronic risk factor. The acute risk factor is defined as the pathophysiological change (vasoconstrictive, hemodynamic, or hemostatic) potentially leading to occlusive coronary thrombosis. It results from a combination of an external stress (physical or mental) and the individual's reactivity to that stress. While the extent of atherosclerosis changes slowly with time (chronic risk factor), hemodynamic, vasoconstrictive, and prothrombotic forces (acute risk factors) may be rapidly generated by external stresses.

Therapeutic Implications

The current state of knowledge of circadian variation of disease onset, and potentially harmful physiological processes, raises the question as to whether pharmacological therapy should be altered to diminish the increased risk of MI in the morning.

For beta blockade, the data appear sufficient to justify selection of an agent that will provide adequate 24-hour protection, particularly in the morning hours. This recommendation is based on substantial evidence, but it is important to note that there has not been, and is unlikely to be, a randomized trial comparing the ability of a long-acting beta blocker versus a shorter acting agent to prevent cardiovascular events.

For aspirin, the issue of morning protection is moot, because a single dose of aspirin provides suppression of morning platelet activity for approximately 3 days. For other agents such as coronary vasodilators and antihypertensive agents, the issue is unresolved. It seems reasonable that pharmacological protection should be provided during the morning hours for patients already receiving anti-ischemic and antihypertensive therapy. However, studies documenting that such a regimen is more likely to prevent MI than a regimen providing less morning protection have not been reported and, to our knowledge, are not in progress or planned.

Since infarcts are more frequent in the morning and can be triggered by exertion, questions about the relative risk of morning exercise versus afternoon or evening exercise have been raised. It is clear that exercise is beneficial in reducing the risk of infarction, and, although theoretical concerns can be raised, there is no evidence that exercise in the morning is more hazardous than exercise at other times of the day. On the contrary, recent data from the Onset Study indicate that the relative risk of exertion in the morning is not substantially greater than the risk of exertion at other times of the day.

The difference between the relative risk versus the absolute risk of experiencing an event during exertion, or any other potential trigger, is also important. While there may exist a sixfold increase in the relative risk of an infarction during exercise in the morning hours, this translates to only a very small increase in absolute risk because baseline risk is low. The absolute risk of having an MI in the hour after exertion might rise from approximately 5/1,000,000 to 5/100,000. Therefore, there is at present *no justification* for a recommendation that afternoon is preferable to morning for exercise.

Future Studies

The primary significance of the recognition of circadian variation of MI onset is the evidence it provides supporting the concept that activities of the patient can *trigger* the onset of MI at any time of day. Triggering is of greater importance than circadian variation of MI because most infarcts do not occur during the morning hours of peak incidence. Understanding triggering could lead to prevention of far more infarctions than reduction of the morning peak of occurrence. To exploit the clues provided by circadian variation of infarction onset, studies ranging from the epidemiological to the molecular level are required.

On the epidemiological level, studies must be conducted in which patients who experience a nonfatal MI are interviewed to determine if the event had an identifiable trigger. Since potentially triggering activities occur frequently without producing an event, the studies must be controlled for the frequency of potential triggers at times when an event did not occur.

The certainty with which an activity can be identified as a trigger will also vary in individual cases. In a patient whose plaque is only slightly vulnerable, the activity required to produce disease onset may be extreme, and the activity can be recognized as a trigger by its intensity. Other features that may aid in the recognition of an activity as a trigger are its occurrence immediately before the event, its ability to produce physiological changes likely to trigger thrombosis, and its absence as part of the patient's routine activity. However, in a patient with an extremely vulnerable plaque, even nonstrenuous, routine, daily activities such as eating a heavy meal may be sufficient to trigger the cascade leading to infarction. In such instances, it may be impossible to identify the triggering activity even though it was present. Thus, the group of patients with *identifiable* triggers will be a subset of those in whom external triggering actually occurred.

The Myocardial Infarction Onset Study, which is supported by the NHLBI, has identified four triggers of onset of infarction: start of activity in the morning, sexual activity, anger, and heavy physical exer-

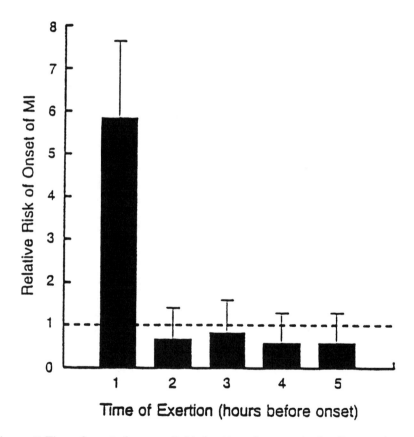

Figure 5: Time of onset of myocardial infarction after an episode of heavy physical exertion. Each of the 5 hours before the onset of myocardial infarction was assessed as an independent hazard period, and exertion during each hour was compared with that during the control period. Only exertion during the hour immediately before the onset of myocardial infarction was associated with an increase in the relative risk, suggesting that the induction time for myocardial infarction is less than 1 hour. The T bars indicate the 95% confidence limits. The dotted line indicates the baseline risk. (Reprinted with permission.[51])

tion[51-53] (Fig. 5). Together these triggers account for approximately 16% of all infarctions, or more than 250,000 events annually. It is of note that the study by Mittleman et al. demonstrated that regular exertion could markedly diminish the likelihood that a single episode of physical exertion would trigger an infarction (Fig. 6).

On the clinical level, increased study of the relationship between daily activities and potentially triggering physiological responses could clarify the manner in which these processes cause disease onset.

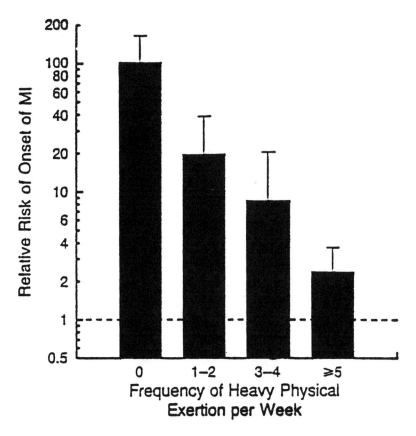

Figure 6: Relative risk of myocardial infarction according to the usual frequency of heavy exertion. Heavy exertion was defined as physical activity at a level of 6 MET or more. The relative risk is shown on a logarithmic scale. Habitually sedentary persons had an extreme relative risk (107), whereas those who reported heavy exertion five or more times per week had a risk only 2.4 times higher than the baseline risk (p <0.001). The T bars indicate the 95% confidence limits. The dotted line indicates the baseline risk. (Reprinted with permission.[51])

On the basic science level, there is a need for complete characterization of the control mechanisms of potentially adverse and beneficial physiological processes. With improved understanding of these mechanisms, clinicians may eventually be able to eliminate unnecessary and potentially detrimental surges in arterial pressure, vasoconstriction, and coagulability that contribute to disease onset, and to increase the activity of potentially beneficial processes such as the fibrinolytic system.

The factors determining plaque vulnerability require further characterization. The reduction in clinical events recently achieved by marked lowering of plasma cholesterol[54] (FATS study) might result not only from a reduced tendency to coronary artery stenosis, but also from a reduction in the formation of lipid pools within plaques that might increase the susceptibility of a plaque to rupture.

Studies of plaque disruption can utilize an atherosclerotic rabbit model of plaque disruption developed in 1964 by Dr. Paris Constantinides, which subsequently received very little attention because of lack of appreciation of the clinical importance of plaque disruption. We have recently reestablished this animal model in our laboratories.[55] Following 8 months on an intermittent cholesterol feeding atherogenic diet, Russell viper venom (a proteolytic procoagulant) and histamine (a pressor in rabbits) are injected into the rabbits to attempt to trigger plaque disruption and thrombosis. Constantinides has documented that this regimen produces localized platelet-rich thrombi overlying disrupted aortic atherosclerotic plaques in approximately 30% of the rabbits. This animal model will be used to test many of the hypotheses generated by the new findings regarding disease onset.

Greater understanding of triggering mechanisms and plaque stabilization should facilitate progress in the prevention of clinical coronary artery disease. The advantage of this approach is that it could lead to effective measures against the majority of coronary heart disease deaths that occur suddenly before any form of therapy can be employed. Even a treatment as effective as thrombolysis prevents less than 3% of the annual deaths attributed to coronary heart disease, since most deaths occur suddenly.[1]

The means of prevention of triggering would not be to eliminate potential triggering activities—an undesirable and unattainable goal—but to design regimens that can be evaluated in randomized studies for their ability to sever the linkage between a potential triggering activity and development of MI.

Acknowledgments: We are grateful for the assistance of Ms. Kathleen Carney in the preparation of the manuscript.

References

1. Muller JE, Tofler GH: Circadian variation and cardiovascular disease. N Engl J Med 1991; 325:1038–1039.
2. Thompson DR, Blandford RL, Sutton TW, et al: Time of onset of chest pain in acute myocardial infarction. Int J Cardiol 1985; 7:139–148.
3. Myers A, Dewar HA: Circumstances attending 100 sudden deaths from coronary artery disease with coroner's necropsies. Br Heart J 1975; 37:1133–1143.
4. World Health Organization: Myocardial infarction community registers: results of a WHO international collaborative study coordinated by the regional office for Europe. 1976; 5:1–23.
5. Muller JE, Stone PH, Turi ZG, et al: Circadian variation in the frequency of onset of acute myocardial infarction. N Engl J Med 1985; 313:1315–1322.
6. Willich SN, Linderer T, Wegscheider K, et al: Increased morning incidence of myocardial infarction in the ISAM Study: absence with prior beta-adrenergic blockade. Circulation 1989; 80:853–858.
7. Goldberg RJ, Brady P, Muller JE, et al: Time of onset of symptoms of acute myocardial infarction. Am J Cardiol 1990; 66:140–144.
8. Willich SN, Lowel H, Lewis M, et al: Association of wake time and onset of myocardial infarction: triggers and mechanisms of myocardial infarction (TRIMM) pilot study. TRIMM Study Group. Circulation 1991; 84:VI62–VI67.
9. Kleiman NS, Schechtman KB, Young PM, et al: Lack of diurnal variation in the onset of non-Q wave infarction. Circulation 1990; 81:548–555.
10. Ogawa H, Misumi I, Masuda T, et al: Difference in plasminogen activator inhibitor-1 (PAI-1) activity between Q wave infarction and non-Q wave infarction. Circulation 1991; 84:II-289.
11. DeWood MA, Spores J, Notske R, et al: Prevalence of total coronary occlusion during the early hours of transmural myocardial infarction. N Engl J Med 1980; 303:897–902.
12. Ambrose JA, Winters SL, Stern A, et al: Angiographic morphology and the pathogenesis of unstable angina pectoris. J Am Coll Cardiol 1985; 5:609–616.
13. Sherman CT, Litvack F, Grundfest W, et al: Coronary angioscopy in patients with unstable angina pectoris. N Engl J Med 1986; 315:913–919.
14. Constantinides P: Plaque fissure in human coronary thrombosis. J Atherosclerosis Res 1966; 1:1–17.
15. Nakagawa S, Hanada Y, Koiwaya Y, et al: Angiographic features in the infarct-related artery after intracoronary urokinase followed by prolonged anticoagulation: role of ruptured atheromatous plaque and adherent thrombus in acute myocardial infarction in vivo. Circulation 1988; 78:1335–1344.
16. Falk E: Plaque rupture with severe pre-existing stenosis precipitating coronary thrombosis: characteristics of coronary atherosclerotic plaques underlying fatal occlusive thrombi. Br Heart J 1983; 50:127–134.
17. Little WC, Constantinescu M, Applegate RJ, et al: Can coronary angiography predict the site of a subsequent myocardial infarction in patients with mild-to-moderate coronary artery disease? Circulation 1988; 78:1157–1166.
18. Brown BG, Gallery CA, Badger RS, et al: Incomplete lysis of thrombus in the moderate underlying atherosclerotic lesion during intracoronary infusion of streptokinase for acute myocardial infarction: quantitative angiographic observations. Circulation 1986; 73:653–661.

19. Haft JI, Haik BJ, Goldstein JE: Catastrophic progression of coronary artery lesions, the common mechanism for coronary disease progression. Circulation 1987; 76:168.

20. Richardson PD, Davies MJ, Born GV: Influence of plaque configuration and stress distribution on fissuring of coronary atherosclerotic plaques. Lancet 1989; 2:941–944.

21. Falk E: Why do plaques rupture? Circulation 1992; 86:III30–42.

22. van der Wal AC, Becker AE, van der Loos CM, et al: Site of intimal rupture or erosion of thrombosed coronary atherosclerotic plaques is characterized by an inflammatory process irrespective of the dominant plaque morphology. Circulation 1994; 89:36–44.

23. Fernandez-Ortiz A, Badimon JJ, Falk E, et al: Characterization of the relative thrombogenicity of atherosclerotic plaque components: implications for consequences of plaque rupture. J Am Coll Cardiol 1994; 23:1562–1569.

24. Alexander RW: Inflammation and coronary artery disease. N Engl J Med 1994; 331:468–469.

25. Tofler GH, Brezinski D, Schafer AI, et al: Concurrent morning increase in platelet aggregability and the risk of myocardial infarction and sudden cardiac death. N Engl J Med 1987; 316:1514–1518.

26. Jimenez AH, Tofler GH, Chen X, et al: Effects of nadolol on hemodynamic and hemostatic responses to potential mental and physical triggers of myocardial infarction in subjects with mild systemic hypertension. Am J Cardiol 1993; 72:47–52.

27. Behar S, Halabi M, Reicher-Reiss H, et al: Circadian variation and possible external triggers of onset of myocardial infarction. SPRINT Study Group. Am J Med 1993; 94:395–400.

28. Weitzman ED, Fukushima D, Nogeire C, et al: Twenty-four hour pattern of the episodic secretion of cortisol in normal subjects. J Clin Endocrinol Metab 1971; 33:14–22.

29. Winther K, Hillegass W, Tofler GH, et al: Effects on platelet aggregation and fibrinolytic activity during upright posture and exercise in healthy men. Am J Cardiol 1992; 70:1051–1055.

30. Peters RW, Zoble RG, Liebson PR, et al: Identification of a secondary peak in myocardial infarction onset 11 to 12 hours after awakening: the Cardiac Arrhythmia Suppression Trial (CAST) experience. J Am Coll Cardiol 1993; 22:998–1003.

31. Somers VK, Dyken ME, Mark AL, et al: Sympathetic-nerve activity during sleep in normal subjects. N Engl J Med 1993; 328:303–307.

32. Hansen O, Johansson BW, Gullberg B: Circadian distribution of onset of acute myocardial infarction in subgroups from analysis of 10,791 patients treated in a single center. Am J Cardiol 1992; 69:1003–1008.

33. Peters RW, Muller JE, Goldstein S, et al: Propranolol and the morning increase in the frequency of sudden cardiac death (BHAT Study). Am J Cardiol 1989; 63:1518–1520.

34. Ridker PM, Manson JE, Buring JE, et al: Circadian variation of acute myocardial infarction and the effect of low-dose aspirin in a randomized trial of physicians. Circulation 1990; 82:897–902.

35. Mulcahy D, Keegan J, Cunningham D, et al: Circadian variation of total ischaemic burden and its alteration with anti-anginal agents. Lancet 1988; 2:755–759.

36. Parker JD, Testa MA, Jimenez AH, et al: Morning increase in ambulatory ischemia in patients with stable coronary artery disease. Importance of physical activity and increased cardiac demand. Circulation 1994; 89: 604–614.

37. Kurnik PB: Circadian variation in the efficacy of t-PA. Circulation 1991; 84:289.

38. Fujita M, Araie E, Yamanishi K, et al: Circadian variation in the success rate of intracoronary thrombolysis for acute myocardial infarction. Am J Cardiol 1993; 71:1369–1371.

39. Hansen O, Johansson BW, Gullberg B: The clinical outcome of acute myocardial infarction is related to the circadian rhythm of myocardial infarction onset. Angiology 1993; 44:509–516.

40. Muller JE, Abela GS, Nesto RW, et al: Triggers, acute risk factors and vulnerable plaques: The lexicon of a new frontier. J Am Coll Cardiol 1994; 23:809–813.

41. Davies MJ: Thrombosis and coronary atherosclerosis. In: Julian DG, Kubler W, Morris RM, Swan HJ, Collen D, Verstraete M, eds. Thrombolysis in cardiovascular disease. New York: Marcel Dekker, 1989.

42. Ip JH, Fuster V, Badimon L, et al: Syndromes of accelerated atherosclerosis: role of vascular injury and smooth muscle cell proliferation. J Am Coll Cardiol 1990; 15:1667–1687.

43. Willerson JT, Campbell WB, Winniford MD, et al: Conversion from chronic to acute coronary artery disease: speculation regarding mechanisms. Am J Cardiol 1984; 54:1349–1354.

44. Coller BS, Folts JD, Smith SR, et al: Abolition of in vivo platelet thrombus formation in primates with monoclonal antibodies to the platelet GPIIb/ IIIa receptor. Correlation with bleeding time, platelet aggregation, and blockade of GPIIb/IIIa receptors. Circulation 1989; 80:1766–1774.

45. Fuster V, Badimon L, Badimon JJ, et al: The pathogenesis of coronary artery disease and the acute coronary syndromes (2). N Engl J Med 1992; 326:310–318.

46. Arbustini E, Grasso M, Diegoli M, et al: Coronary thrombosis in non-cardiac death. Coronary Artery Disease 1993; 4:751–759.

47. Belch JJ, McArdle BM, Burns P, et al: The effects of acute smoking on platelet behaviour, fibrinolysis and haemorheology in habitual smokers. Thromb Haemost 1984; 51:6–8.

48. Egan B, Schork N, Panis R, et al: Vascular structure enhances regional resistance responses in mild essential hypertension. J Hypertension 1988; 6:41–48.

49. Ludmer PL, Selwyn AP, Shook TL, et al: Paradoxical vasoconstriction induced by acetylcholine in atherosclerotic coronary arteries. N Engl J Med 1986; 315:1046–1051.

50. Khanna PK, Seth HN, Balasubramanian V, et al: Effect of submaximal exercise on fibrinolytic activity in ischaemic heart disease. Br Heart J 1975; 37:1273–1276.

51. Mittleman MA, Maclure M, Tofler GH, et al, for the Determinants of Myocardial Infarction Onset Study Investigators: Triggering of acute myocardial infarction by heavy exertion: Protection against triggering by regular exertion. N Engl J Med 1993; 329:1677–1683.

52. Muller JE, Maclure M, Mittleman MA, et al: Risk of myocardial infarction doubles in the two hours after sexual activity, but absolute risk remains low. Circulation 1993; 88:509.
53. Mittleman MA, Maclure M, Sherwood JB, et al: Triggering of myocardial infarction onset by episodes of anger. Circulation 1994; 936.
54. Brown G, Albers JJ, Fisher LD, et al: Regression of coronary artery disease as a result of intensive lipid-lowering therapy in men with high levels of apolipoprotein B. N Engl J Med 1990; 323:1289–1298.
55. Abela GS, Picon PD, Friedl S, et al: Triggering of plaque disruption and arterial thrombosis in an atherosclerotic rabbit model. Circulation 1995; 91:776–784.

7

Cardiac Arrhythmias and Circadian Changes

Simon Chakko, MD, Robert J. Myerburg, MD

Introduction

Recent studies have demonstrated that sudden cardiac death, myocardial infarction, and transient myocardial ischemia distribute in time in a circadian pattern. These findings, which are discussed in detail in other chapters of this book, raise the possibility that the occurrence of cardiac arrhythmias may also be expressed in a circadian pattern. In this chapter, we describe the temporal patterns in onset of arrhythmias and discuss the underlying mechanisms and difficulties of studying such patterns.

Complexities of Studying Circadian Patterns of Arrhythmias

Investigating the circadian pattern of arrhythmias is a difficult task because arrhythmogenesis is a complex phenomenon. It is important to understand that both structural abnormalities and transient functional factors commonly interact in the pathogenesis of arrhythmias.[1] An underlying structural abnormality (e.g., coronary artery disease, cardiomyopathy, Wolff-Parkinson-White syndrome, etc.) may cause an arrhythmia independently or may interact with transient functional factors (e.g., hemodynamic, neurohormonal, electrolyte, and

From: Deedwania PC (ed): *Circadian Rhythms of Cardiovascular Disorders.*
©Futura Publishing Co., Inc., Armonk, NY, 1997.

respiratory abnormalities) to initiate an arrhythmia. These transient functional factors, if severe, are also capable of causing arrhythmias in the absence of structural abnormalities. In the study of circadian patterns of arrhythmias, primary and secondary arrhythmias must be distinguished.[1] An arrhythmia that results from an electrophysiological disturbance caused by a disease process independent of coexistent hemodynamic factors is defined as primary arrhythmia. An arrhythmia that results from an electrical disturbance initiated by hemodynamic deterioration or metabolic abnormality is a secondary arrhythmia. There is no uniform hypothesis regarding mechanisms by which these elements interact to lead to the final pathway of lethal arrhythmias.[2,3] The onsets of acute myocardial infarction and other acute ischemic events have a circadian rhythm with a primary peak incidence in the morning.[4,5] Primary ventricular fibrillation is a common early complication of acute ischemic events.[1] Thus, a circadian pattern in the onset of ventricular fibrillation can be expected if patients with acute myocardial infarction are studied. However, it is very unlikely that such a pattern will emerge if secondary ventricular fibrillation is studied.

Studying circadian patterns of lethal ventricular arrhythmias raises an additional problem. In epidemiological studies in which most investigated deaths occur outside the hospital setting and are not witnessed by medically trained persons, it is difficult to determine whether the death was caused by an arrhythmia or by circulatory failure. A terminal acute illness of short duration (less than 1 hour) is a strong indicator of arrhythmic death. A longer terminal illness may indicate death due to either arrhythmia or circulatory failure.[6] Death certificates and emergency room records may not provide sufficient information to determine whether the cause of death was an arrhythmia. Many sudden cardiac deaths are unwitnessed and it may be impossible to determine the exact time of death. Unwitnessed deaths that are more likely to occur during the night may be excluded from analysis, thereby artificially lowering the frequency of events during the night.

Many extraneous factors, e.g., caffeine, alcohol, sympathomimetic agents in nonprescription drugs, and illicit drugs such as cocaine and amphetamines, may trigger and aggravate arrhythmias.[7] Failure to recognize these confounding factors may result in inappropriate conclusions. Autonomic activity has an intrinsic circadian rhythm and likely contributes to the pattern of circadian variation in the onset of cardiac events. As discussed below, in some disease states, e.g., diabetes, this circadian rhythm is lost.[8] In such subgroups, the circadian patterns in the onset of cardiac diseases have been reported to be altered.[9]

Ambulatory ECG monitoring is commonly used to evaluate circadian patterns of arrhythmia. Morganroth et al.[10] evaluated the variations in the frequency of ventricular premature depolarizations (VPDs) with three consecutive 24-hour ECG monitorings. The extent of spontaneous variation in arrhythmia frequency that occurred in individual patients from day-to-day was 23%, between 8-hour periods was 29%, and hour-to-hour was 48%. As shown in Figure 1, the circadian pattern of nocturnal decrease in VPDs may not be seen every day and the number of VPDs varies greatly. Day-to-day variation in VPD frequency is less marked when the frequency of ventricular arrhythmia is high. Michelson[11] et al. applied statistical methods to determine the change in frequency of complex ventricular arrhythmias that is necessary to exceed that attributable to spontaneous variation alone at the $p < 0.05$ level. No such data are available to determine the statistical significance of diurnal variations in arrhythmia occurrence.

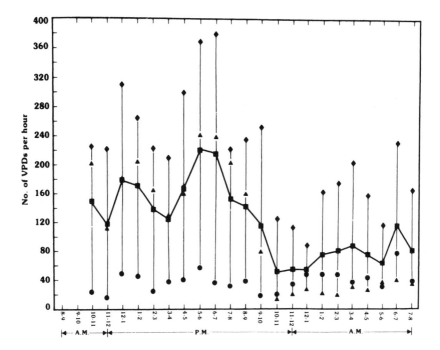

Figure 1: Variation in the number of ventricular premature depolarizations (VPDs) per hour on 3 consecutive days (day 1: diamonds; day 2: triangles; day 3: circles; average of 3 days: squares) in a patient with no known heart disease. Note that a pattern of decreasing number of VPDs during sleep with an increase during the morning hours is clear in the averaged data and on days 1 and 2. No such pattern is noted on day 3. (Reproduced with permission from the American Heart Association.[10])

Mechanisms of Circadian
of Arrhythmias

Muller et al.[12] reviewed the circadian rhythm of the triggering factors of acute cardiovascular disease. A number of physiological processes that may lead to atherosclerotic plaque rupture and coronary thrombosis (e.g., increased sympathetic tone, hypercoagulable state, and coronary vasoconstriction) have a circadian rhythm that is accentuated in the morning. The frequencies of arrhythmogenic events, such as the onset of transient myocardial ischemia or acute myocardial infarction, show a parallel increase in the morning, as does sudden cardiac death.

The autonomic nervous system plays a significant role in modulating cardiac arrhythmias.[13] Increased sympathetic neural activity accelerates heart rate, favors spontaneous depolarization, shortens the ventricular effective refractory period, and decreases the threshold for ventricular fibrillation.[14] In contrast, increased parasympathetic activity slows the heart rate, decreases atrioventricular nodal conduction, and in the presence of baseline sympathetic neural activity, increases both the ventricular refractory period and the ventricular fibrillation threshold.[15] Heart rate variability obtained from 24-hour ambulatory ECG monitoring is a reliable noninvasive measure of autonomic tone fluctuations.[14,16] Standard deviation of the R-R interval and percentage of adjacent R-R cycles with differences greater than 50 ms (PNN-50) are measures of vagal tone. Spontaneous episodes of ventricular tachycardia are often preceded by changes in heart rate variability.[17] Decreased heart rate variability is a strong and independent predictor of mortality after acute myocardial infarction.[18] Heart rate variability, which is predominantly influenced by the cardiac vagal activity, has a reproducible circadian rhythm.[19] It increases during sleep and decreases abruptly during the hours after waking. Heart rate, which is influenced by vagal and sympathetic balance, decreases during sleep and increases upon awakening. This circadian pattern is lost in many diseases (e.g., diabetes,[9] hypertension,[20] congestive heart failure,[21] and myocardial infarction[22]). Figures 2 and 3 show the circadian pattern of heart rate and PNN-50, a measure of parasympathetic tone. Note that the heart rate decreases at night and increases during the day (Fig. 2). Figure 3 demonstrates that parasympathetic tone increases at night and decreases during the day; note that this circadian pattern is blunted in hypertensive patients with left ventricular hypertrophy.

Prolonged QT interval is associated with ventricular arrhythmias. Diurnal variation in the QTc interval (QT corrected to heart rate by Bazett's formula) has been reported among patients with normally innervated hearts.[23] This diurnal variation is blunted in heart transplant

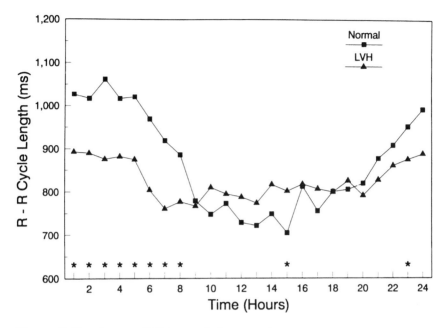

Figure 2: Hourly mean R-R intervals in normal controls and hypertensive patients with left ventricular hypertrophy. R-R intervals increase at night and decrease during the day. This circadian pattern is blunted in hypertensive patients with left ventricular hypertrophy. (Reproduced with permission.[20])

recipients and in insulin-dependent diabetics. Changes in sympathovagal balance measured by heart rate variability have been demonstrated to be responsible for the diurnal variation in QT interval.[24]

During sleep, sympathetic neural activity decreases and parasympathetic activity predominates[14] and bradyarrhythmias are common. Marked sinus bradycardia with junctional escape rhythm, Wenkebach type of second degree AV block, and very slow ventricular response to atrial fibrillation are common physiological phenomena during sleep. Ambulatory ECG monitoring of 50 healthy young men revealed that during sleep 28% had sinus pauses of more than 1.75 seconds and 6% had Wenkebach type of second degree AV block.[25] In patients with obstructive sleep apnea, cyclical variation in heart rate (sinus bradytachyarrhythmia) during sleep is typical. The bradycardia that coincides with apnea can be abolished with atropine. Sudden increase in the heart rate occurs following the termination of apnea.[14] Sinus pauses, second degree atrioventricular block, increase in the number of premature ventricular contractions, and ventricular tachycardia have been reported in obstructive sleep apnea.[26] Although the prevalence of arrhythmias in sleep apnea is not different from that seen in healthy

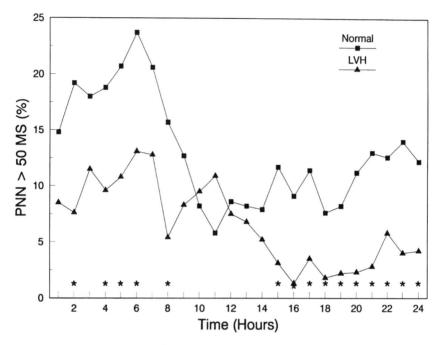

Figure 3: PNN-50 (percentage of adjacent R-R cycles with differences that exceed 50 ms), a measure of vagal tone in normal controls and hypertensive patients with left ventricular hypertrophy. Note the circadian pattern in normals. (Reproduced with permission.[20])

young men,[14] these arrhythmias are associated with hypoxia and can be eliminated by the correction of hypoxia.[14,26] Nocturnal arrhythmias associated with hypoxia are also seen in obstructive lung disease and respond to correction of hypoxia.[27]

Frequency of ventricular arrhythmias appears to decrease during sleep. Lown et al.[28] compared the incidence of ventricular premature beats (VPBs) during sleep to the awake state in 54 patients. During sleep, the number of VPBs decreased by 50% in 22 patients and by 25–50% in an additional 13 patients. While a decrease in VPBs was evident in this study during sleep, atrial premature beats increased. Raeder et al.[29] reported that the nocturnal decrease in VPBs was reproducible in two successive 24-hour monitorings. The decrease in VPBs is probably related to the decrease in sympathetic tone during sleep. Simultaneous ECG and electroencephalogram monitoring revealed that the decrease in VPBs correlated more closely with the change in heart rate than with the level of arousal monitored by electroencephalography.[30] Two studies reported that the incidence of VPBs did not decrease during sleep.[31,32] However, one of these studies was performed in the coronary care unit where sleep patterns may be altered[31] and the other studied patients with recent myocardial infarction whose circadian rhythm of autonomic tone may be blunted.[32] Circadian rhythms of physiological parameters may be altered in subjects who have an altered sleep pattern. Mulcahy et al.[33] studied the effect of an afternoon nap on heart rate and blood pressure. There was a decrease in both of these parameters during a nap with an abrupt increase upon awakening. Thus sleep appears to have a protective effect regardless of when it occurs.

Cinca et al.[34] assessed the circadian variations in the electrical properties of the human heart by sequential bedside electrophysiological testing. It was noted that between 12:00 midnight and 7:00 AM, there was a significant lengthening of sinus node rate, sinus node recovery time, QT interval, and the effective refractory period of the atria, AV node, and right ventricle. Thus, conventional electrophysiological parameters are subject to daily variability. AV nodal and myocardial refractoriness follow a circadian rhythm with an acrophase between 12:00 midnight and 7:00 AM. However, clinical significance of these findings is unclear since the time of the day when electrophysiological studies are performed does not appear to affect the test results. McClelland et al.[35] reported the results of two drug-free ventricular stimulation studies performed 4 to 28 hours apart in each of the 162 patients studied for evaluation of serious ventricular arrhythmias. Changes in the rate and duration of induced arrhythmia and the number of extrastimuli required to induce arrhythmia during the two tests were compared. No significant circadian variation was found in these electrophysiological measures. It should be noted that none of the studies were performed between midnight and 7:00 AM. Moreover, the study included only hospitalized patients in whom the circadian rhythms may be blunted.

Supraventricular Tachycardias

Irwin et al.[36] studied the circadian pattern of occurrence of symptomatic paroxysmal supraventricular tachycardia in 52 patients who were not receiving treatment. The tachycardias were AV nodal reentrant tachycardia or AV reciprocating tachycardia. Patients transmitted their ECGs by telephone when symptoms occurred. Thus, asymptomatic tachycardias were not studied. Symptomatic tachycardias had a circadian pattern with a peak incidence at 4 PM, with a corresponding minimum incidence at 4 AM; patients were five times more likely to have tachycardia in the afternoon than in the morning (Fig. 4). The

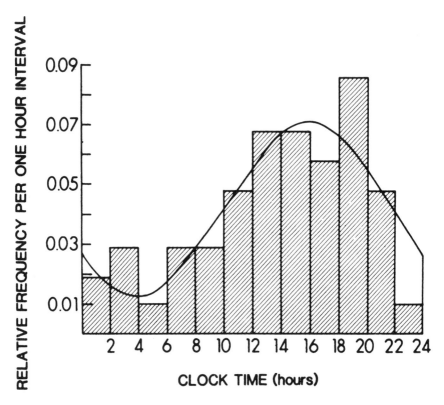

Figure 4: Relative frequency histogram for the occurrence of paroxysmal supraventricular tachycardia during 24 hours. Note the peak incidence at 4:00 PM and a minimum incidence at 4:00 AM. (Reproduced with permission from the American Heart Association.[36])

time of the second episode of tachycardia was independent of the time of the first episode. A likely explanation for this observed circadian rhythm is fluctuations in autonomic tone. The higher incidence of paroxysmal supraventricular tachycardia during the daytime hours raises the question of whether antiarrhythmic therapy should be more intense during these hours. Pritchett et al.[37] documented symptomatic paroxysmal supraventricular tachycardias by telephone transmission of ECG in patients who were being treated with verapamil or diltiazem. There was no predilection for tachycardia to occur late in a dosing interval when the plasma drug levels were presumed to be the lowest. Episodes of tachycardia were uniformly distributed throughout the wakeful hours, between 9 AM and 12 midnight. No tachycardia was noted between 12 midnight to 9 AM.

Kupari et al.[38] studied the circadian rhythm of sustained supraventricular tachycardias using a different methodology. They studied 209 patients who presented to the emergency room with a supraventricular arrhythmia (including atrial flutter and atrial fibrillation) and were able to tell the exact time when their symptoms had begun. There were two peaks in the frequency of onset of supraventricular tachycardias with one in the morning between 6:01 AM and noon and another in the evening between 6:01 PM and midnight and a trough at night. Neither the etiology of the arrhythmia nor preceding alcohol consumption appeared to modify this tendency. However, the two-peaking variation was more characteristic of atrial fibrillation. Among 50 patients who were using beta-adrenergic blocking drugs, the morning surge of arrhythmia was absent and a higher incidence was noted at night between midnight and 6:00 AM. Such a blunting of circadian rhythm with an absent morning surge also has been reported with the onset of myocardial infarction among patients receiving beta-adrenergic blockers.[39]

Rostagno et al.[40] evaluated the time of onset of symptomatic atrial fibrillation and paroxysmal supraventricular tachycardia among patients who were treated by a mobile coronary care unit in Florence, Italy. Among the 726 patients with atrial fibrillation, peak incidence was between midnight and 2:00 AM; secondary peaks were found in the morning between 8:00 AM and 9:00 AM and in the afternoon between 2:00 PM and 4:00 PM. No significant difference in time of occurrence was found between lone atrial fibrillation and atrial fibrillation associated with organic heart disease. In contrast, among the 348 patients with paroxysmal supraventricular tachycardia, maximal incidence was during the daytime with a marked decrease between midnight and 8:00 AM. The investigators postulated that increased vagal drive at night might explain the higher incidence of atrial fibrillation, since, experimentally, atrial fibrillation can be induced by the administration of acetylcholine.[40]

The studies of circadian variations of supraventricular arrhythmias have obvious drawbacks. Each investigated only symptomatic arrhythmias. Only two studies had large numbers of patients[38,40] and they evaluated patients who were treated in an emergency room or mobile coronary care unit. They were conducted in different countries where thresholds for seeking emergency care may be different. Patients with mildly symptomatic supraventricular arrhythmias may not seek emergency therapy. However, all three studies reported a low incidence of paroxysmal supraventricular tachycardia at night. The contradictory results for the peak time of onset of atrial fibrillation are unexplained, but may result from the differences in study populations and methodologies. Circadian fluctuation of ventricular response to atrial fibrillation has been studied using ambulatory ECG monitoring performed outside the hospital.[41] The ventricular rate revealed a prominent circadian pattern with an estimated peak at 1:01 PM and a trough at 3:40 AM. The amplitude of this fluctuation was 22%. The likely explanation for this pattern is the increased vagal tone during sleep causing an increase in the refractory period of the AV node.

Ventricular Arrhythmias

Ventricular Premature Beats (VPBs)

Diurnal rhythm of VPBs has been evaluated by many investigators.[28–32] Patients included in these studies often had coronary artery disease and high density of VPBs. Frequency of ventricular arrhythmias appears to decrease at night during sleep. Lown et al.[28] compared the incidence of VPBs during sleep to the awake state in 54 patients. During sleep, the number of VPBs decreased by 50% in 22 patients and by 25–50% in an additional 13 patients. Steinbach et al.[42] reported a similar nocturnal decrease in VPBs in the majority of the 77 patients studied, with a minimum number of VPBs recorded between midnight and 2:00 AM; maximum VPB frequency was widely dispersed throughout the day without a clear peak. Other studies designed to evaluate day-to-day variations in the frequency of VPBs also have reported a circadian pattern of VPBs with a nocturnal decrease[10,11,29] (Fig. 1).

Patients with hypertension and ECG abnormalities are at increased risk for sudden death. Seigel et al.[43] reported the circadian variation of ventricular arrhythmias in 199 men with hypertension and ECG abnormalities; patients with myocardial infarction, angina, and electrolyte abnormalities were excluded.

The VPB frequency in the 24-hour ambulatory ECG monitoring was small and the majority of patients had less than 10 VPBs per hour.

Ventricular arrhythmias were classified by 6-hour time intervals. The interval from 6:00 AM to noon revealed a higher prevalence of complex or frequent VPBs and a higher mean number of VPBs per hour when compared to the interval from midnight to 6:00 AM. VPBs during the other two 6-hour periods were intermediate in frequency. This study is interesting since a circadian rhythm of VPBs was demonstrated in the absence of high frequency of arrhythmia.

It is well known that the frequency of VPBs is highly variable from day to day and hour to hour, especially when the frequency of VPBs is small.[11] Thus it is important to document that the observed circadian rhythm is reproducible. Raeder et al.[29] reported that the nocturnal decrease in VPBs was reproducible in two successive 24-hour monitorings. Lanza et al.[44] evaluated the reproducibility of circadian rhythm in 19 patients with coronary artery disease and 19 with structurally normal hearts using two 24-hour ambulatory ECG monitorings. A significant and similar circadian rhythm of VPBs was found in the total sample with two main peaks, the first in the late morning and the second during the afternoon hours. Patients with coronary artery disease as well as those with structurally normal hearts had similar circadian rhythms. However, among the 38 patients studied, only 18 (47%) had significant individual circadian rhythm of VPBs on both days. When patients with and without circadian rhythm were compared, no significant difference was found in gender, incidence of coronary artery disease, or VPB frequency.

The nocturnal decrease and the morning increase in VPBs is probably related to the alterations in sympathetic tone. Simultaneous ECG and electroencephalogram monitoring revealed that the decrease in VPBs correlated more closely with the change in heart rate than the level of arousal.[30] Two studies reported that the incidence of VPBs did not decrease during sleep.[31,32] However, one of these studies was performed in the coronary care unit where sleep patterns may be altered[31] and the other studied patients with recent myocardial infarction whose circadian rhythm of autonomic tone may be blunted.[32] Mulrow et al.[45] evaluated the circadian rhythm of VPBs in 10 patients with hypertrophic cardiomyopathy. A peak frequency was noted at night in five patients and an afternoon peak in five patients. The finding of a nocturnal peak in half of the patients studied is intriguing, but has to be confirmed in larger numbers of patients.

Clinical observations suggest that beta-adrenergic blocking agents can modify the circadian occurrence of a variety of ischemic events.[46] The effects of propranolol on ventricular arrhythmias were studied by reviewing the data from the BHAT trial, which was a multicenter, randomized, double-blind, placebo-controlled clinical trial designed to evaluate the effectiveness of propranolol in reducing the mortality rate in patients with recent myocardial infarction.[47] A random sample of 25% of patients enrolled in that study underwent 24-hour ambulatory ECG monitoring at baseline and after 6 weeks of therapy. There was no significant decrease in the frequency of VPBs. However, propranolol eradicated the increase in VPBs that occurred with awakening and persisted throughout the day [46, 47] (Fig. 5). Similar results were reported by Nademanee et al.[48] who studied the effect of placebo and nadolol on the circadian rhythm of VPBs and ventricular tachycardia. An increase in VPBs noted during the daytime was abolished with nadolol but was unchanged with placebo.

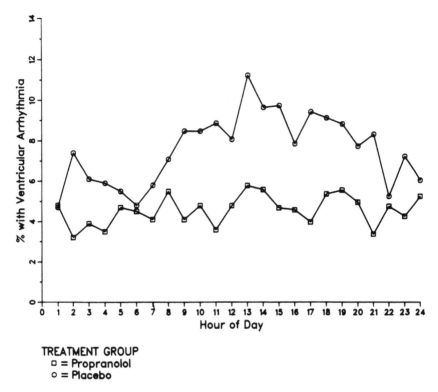

TREATMENT GROUP
□ = Propranolol
○ = Placebo

Figure 5: Effect of propranolol (squares) compared to placebo (circles) on ventricular arrhythmias in the BHAT study. Proportion of patients having ventricular arrhythmia (>10 VPBs and at least one run of ventricular tachycardia or a paired VPB) is shown. (Reproduced with permission from the American Heart Association.[47])

Ventricular Tachycardia

Twidale et al.[49] studied the time of onset of ventricular tachycardia (VT) in 68 patients. In 53 patients, the VT was not associated with an acute myocardial infarction and was documented by 12-lead ECG; in the remaining 15 patients, there was a history of syncope or presyncope of unknown etiology and sustained monomorphic VT was induced at electrophysiological study. None of the patients had a previous history of sustained VT. The time of occurrence of VT was determined by interviewing the patient and family members and grouped into hourly intervals. The peak incidence for the onset of VT was between 10:00 AM and 12:00 noon (Fig. 6). This pattern is similar to that reported for acute myocardial infarction, sudden cardiac death, and transient myocardial ischemia.

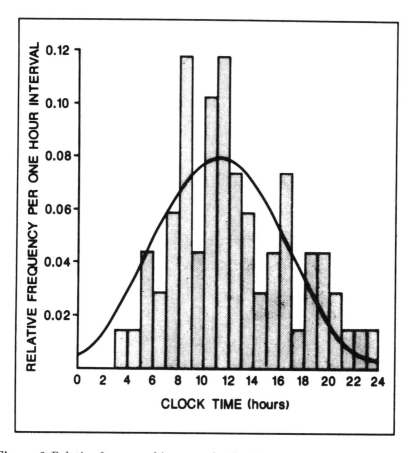

Figure 6: Relative frequency histogram for the time of onset of sustained ventricular tachycardia during the day. The peak incidence is between 10:00 AM and 12:00 noon. (Reproduced with permission.[49])

Among 1,057 patients who had 24-hour Holter monitoring performed after a myocardial infarction, Lucente et al. identified 94 patients who were not receiving beta-adrenergic blockers or other antiarrhythmic drugs and had 1–10 episodes of VT (>3 beats) in 24 hours.[50] Forty-seven (50%) patients had a recent acute myocardial infarction and the rest had a myocardial infarction in the past. All VTs were nonsustained. A significant circadian rhythm in the frequency of VT was noted. VT occurred mostly in the waking hours with two peaks: a primary one in the late morning and a secondary one in the afternoon. A late morning peak prevailed among the patients with remote myocardial infarction and an afternoon peak prevailed among the patients with recent myocardial infarction. However, the significance of this finding is difficult to determine since the patients with recent myocardial infarction were studied in the hospital and the others were studied at home.

Sudden Cardiac Death

Among patients who suffer cardiac arrest or sudden cardiac death, two types of arrhythmias are common: ventricular tachyarrhythmias (VT or ventricular fibrillation) and bradyarrhythmias and asystole.[2] Epidemiological studies that reviewed mortality records have demonstrated an increase in the incidence of sudden cardiac death during the morning hours.[51] From a retrospective review of 26,798 death certificates, Muller et al. identified 2,023 individuals who met the definition for sudden death.[52] The data revealed a prominent circadian variation in the occurrence of sudden death with a primary peak from 7:00 AM to 11:00 AM. This study was based on death certificates and has obvious limitations. In the Framingham heart study,[51,53] all data regarding the death of an individual are reviewed by a Mortality Review Committee which determines the cause of death. Among 2,458 deaths, 264 definite sudden cardiac deaths were identified, and in 59% of these the precise time of death could be identified. There was a sharp increase in the frequency of sudden cardiac death from 6:00 AM to 9:00 AM followed by an equal distribution of sudden cardiac deaths during the rest of the day. This circadian variation was not altered by gender or age. The circadian rhythm in the occurrence of sudden cardiac death is discussed in detail in the next chapter.

Unresolved Issues

It has been demonstrated that a considerable number of patients with cardiac arrhythmias demonstrate a circadian rhythm to their arrhythmia occurrence. However, at present it is not clear whether recognizing such circadian rhythms is of any value. Studies are needed to determine the effectiveness of antiarrhythmic therapy tailored to be more intense during the hours when the frequency of arrhythmias increases. It remains to be seen whether such tailored therapy will reduce the total number of arrhythmic events or simply shift the time of onset of arrhythmia.

References

1. Myerburg RJ, Kessler KM, Castellanos A: Recognition, clinical assessment and management of arrhythmias and conduction disturbances. In: Schlant RC, Alexander RW (eds), Hurst's The Heart, ed 8, New York, McGraw-Hill Inc, 1994; pp 705–774.
2. Myerburg RJ, Castellanos A: Cardiac arrest and sudden cardiac death. In: Braunwald E (ed), Heart Disease, ed 4, Philadelphia, W.B. Saunders Company, 1992; pp 756–789.

3. Myerburg RJ, Kessler KM, Bassett AL, Castellanos A: A biological approach to sudden cardiac death: structure, function and cause. Am J Cardiol 1989; 63:1512–1516.

4. Muller JE, Stone PH, Turi ZG, Rutherford JD, Czeisler CA, Parker C, Poole WK, Passamani E, Roberts R, Robertson T, Sobel BE, Willerson JT, Braunwald E, and the MILIS Study Group: Circadian variation in the frequency of onset of acute myocardial infarction. N Eng J Med 1985; 313:1315–1322.

5. Rocco MB, Barry J, Campbell S, Nabel E, Cook F, Goldman L, Selwyn AP: Circadian variation of transient myocardial ischemia in patients with coronary artery disease. Circulation 1987; 75:359–400.

6. Hinkle LE, Thaler HT: Clinical classification of cardiac deaths. Circulation 1982; 65:457–464.

7. Chakko S, Kessler KM: Recognition and management of arrhythmias. Curr Probl Cardiol 1995; 20:53–120.

8. Bernardi L, Ricordi L, Lazzari P, Solda P, Calciati A, Ferrari MR, Vandea I, Finardi G, Fratino P: Impaired circadian modulation of sympathovagal activity in diabetes: a possible explanation for altered temporal onset of cardiovascular disease. Circulation 1992; 86:1443–1452.

9. Gilpin EA, Hjalmarson A, Ross J: Subgroups of patients with atypical circadian patterns of symptoms onset in acute myocardial infarction. Am J Cardiol 1990; 66:7G–11G.

10. Morganroth J, Michelson EL, Horowitz LN, Josephson ME, Pearlman AS, Dunkman B: Limitations of routine long-term electrocardiographic monitoring to assess ventricular ectopic frequency. Circulation 1978; 58: 408–414.

11. Michelson EL, Morganroth J: Spontaneous variability of complex ventricular arrhythmias detected by long-term electrocardiographic recording. Circulation 1980; 61:690–695.

12. Muller JE, Tofler GH, Stone PH: Circadian variation and triggers of onset of acute cardiovascular disease. Circulation 1989; 79:733–743.

13. Zipes DP, Miyazaki T: The autonomic nervous system and the heart: basis for understanding interactions and effects on arrhythmia development. In: Zipes DP, Jalife J (eds), Cardiac Electrophysiology from Cell to Bedside, Philadelphia, W.B. Saunders, 1990; pp 312–330.

14. Shepard JW: Hypertension, cardiac arrhythmias, myocardial infarction and stroke in relation to obstructive sleep apnea. Clin Chest Med 1992; 43:437–458.

15. Lown B, Verrier RL: Neural activity and ventricular fibrillation. N Engl J Med 1976; 294:1165–1170.

16. Hayano J, Sakakibara Y, Yamada A, Yamada M, Mukai S, Fujinami T, Yokoyama K, Watanabe Y, Takata K: Accuracy of assessment of cardiac vagal tone by heart rate variability in normal subjects. Am J Cardiol 1991; 67:199–204.

17. Huikuri HV, Valkama JO, Airaksinen J, Seppanen T, Kessler KM, Takkunen JT, Myerburg RJ: Frequency domain measures of heart rate variability before the onset of nonsustained and sustained ventricular tachycardia in patients with coronary artery disease. Circulation 1993; 87: 1220–1228.

18. Kleiger RE, Miller JP, Bigger JT, Moss AJ and the Multicenter Post-infarction Research Group: Decreased heart rate variability and its association with increased mortality after acute myocardial infarction. Am J Cardiol 1987; 59:256–262.

19. Huikuri HV, Kessler KM, Terracall E, Castellanos A, Linnaluoto MK, Myerburg RJ: Reproducibility and circadian rhythm of heart rate variability in healthy subjects. Am J Cardiol 1990; 65:391–393.

20. Chakko S, Mulingtapang RF, Huikuri HV, Kessler KM, Materson BJ, Myerburg RJ: Alterations in heart rate variability and its circadian rhythm in hypertensive patients with left ventricular hypertrophy free of coronary artery disease. Am Heart J 1993; 126:1364–1372.

21. Malliani A, Pagani M, Lombardi F, Ceruti S: Cardiovascular neural regulation explored in the frequency domain. Circulation 1991; 84:482–492.

22. Malik M, Farrell T, Camm AJ: Circadian rhythm of heart rate variability after myocardial infarction and its influence on the prognostic value of heart rate variability. Am J Cardiol 1990; 66:1049–1054.

23. Bexton RS, Vallin HO, Camm AJ: Diurnal variation of the QT interval-influence of the autonomic nervous system. Br Heart J 1986; 55:253–258.

24. Murakawa Y, Inoue H, Nozaki A, Sugimoto T: Role of sympathovagal interaction in diurnal variation of QT interval. Am J Cardiol 1992; 69:339–343.

25. Brodsky M, Wu D, Denes P, Kanakis C, Rosen KM: Arrhythmias documented by 24 hour continuous electrocardiographic monitoring in 50 male medical students without apparent heart disease. Am J Cardiol 1977; 39:390–395.

26. Guilleminault C, Connolly SJ, Winkle RA: Cardiac arrhythmia and conduction disturbances during sleep in 400 patients with sleep apnea syndrome. Am J Cardiol 1983; 52:490–494.

27. Flick MR, Block J: Nocturnal vs diurnal cardiac arrhythmias in patients with chronic obstructive lung disease. Chest 1979; 75:8–11.

28. Lown B, Tykocinski M, Garfein A, Brooks P: Sleep and ventricular premature beats. Circulation 1973; 48:691–701.

29. Raeder EA, Hohnloser SH, Graboys TB, Podrid PJ, Lampert S, Lown B: Spontaneous variability and circadian distribution of ectopic activity in patients with malignant ventricular arrhythmia. J Am Coll Cardiol 1988; 12:656–661.

30. Pickering TG, Johnston J, Honour AJ: Comparison of the effects of sleep, exercise and autonomic drugs on ventricular extrasystoles, using ambulatory monitoring of electrocardiogram and electroencephalogram. Am J Med 1978; 65:575–583.

31. Smith R, Johnson L, Rothfeld D, Zir L: Sleep and cardiac arrhythmias. Arch Intern Med 1972; 130:751–753.

32. Morrison GW, Kumar EB, Portal RW, Aber C: Cardiac arrhythmias 48 hours before, during, and 48 hours after discharge from hospital following acute myocardial infarction. Br Heart J 1981; 45:500–511.

33. Mulcahy D, Wright C, Sparrow J, Cunningham D, Curcher D, Purcell H, Fox K: Heart rate and blood pressure consequences of an afternoon SIESTA (Snooze-Induced Excitation of Sympathetic Triggered Activity). Am J Cardiol 1993; 71:611–614.

34. Cinca J, Moya A, Figueras J, Roma F, Ruis J: Circadian variations in the electrical properties of the human heart assessed by sequential bedside electrophysiological testing. Am Heart J 1986; 112:315–321.

35. McClelland J, Halperin B, Cutler J, Kudenchuk P, Kron J, McAnulty: Circadian variation in ventricular electrical instability associated with coronary artery disease. Am J Cardiol 1990; 65:1351–1357.

36. Irwin JM, McCarthy EA, Wilkinson WE, Pritchett ELC: Circadian occur-

rence of symptomatic paroxysmal supraventricular tachycardia in untreated patients. Circulation 1988; 77:298–300.

37. Pritchett ELC, Smith MS, McCarthy EA, Lee KL: The spontaneous occurrence of paroxysmal supraventricular tachycardia. Circulation 1984; 70:1–6.

38. Kupari M, Koskinen P, Leinonen H: Double peaking circadian variation in the occurrence of sustained supraventricular tachyarrhythmias. Am Heart J 1990; 120:1364–1369.

39. Willich SN, Linderer T, Wegscheider K, Leizorovicz A, Alamercery I, Schroder R, and the ISAM Study Group: Increased morning incidence of myocardial infarction in the ISAM Study: absence with prior beta-adrenergic blockade. Circulation 1989; 80:853–858.

40. Rostagno C, Taddei T, Paladini B, Modesti PA, Utari P, Bertini G: The onset of symptomatic atrial fibrillation and paroxysmal supraventricular tachycardia is characterized by different circadian rhythms. Am J Cardiol 1993; 71:453–455.

41. Raeder E: Circadian fluctuations in ventricular response to atrial fibrillation. Am J Cardiol 1990; 66:1013–1016.

42. Steinbach K, Weber GH, Joskowicz GJ, Kaindl F: Frequency and variability of ventricular premature contractions: the influence of heart rate and circadian rhythms. Pace 1982; 5:38–51.

43. Seigel D, Black D, Seeley DG, Hulley SB: Circadian variation in ventricular arrhythmias in hypertensive men. Am J Cardiol 1992; 69:344–347.

44. Lanza GA, Cortellessa MC, Rebuzzi AG, Scabbia EV, Costalunga A, Tamburi S, Lucente M, Manzoli U: Reproducibility in circadian rhythm of ventricular premature complexes. Am J Cardiol 1990; 66:1099–1106.

45. Mulrow JP, Healy MA, McKenna WJ: Variability of ventricular arrhythmias in hypertrophic cardiomyopathy and implications for treatment. Am J Cardiol 1986; 58:615–618.

46. Goldstein S: Effect of beta-adrenergic blocking agents on the circadian occurrence of ischemic cardiovascular events. Am J Cardiol 1990; 66:63G–65G.

47. Lichstein E, Morganroth J, Harrist R, Hubble E, for the BHAT Study Group: Effect of propranolol on ventricular arrhythmia. The Beta-blocker Heart Attack Trial experience. Circulation 1983; 67:I5–I10.

48. Nademanee K, Olukotun AY, Robertson HA, Harwood BJ, Singh BN: Effect of beta-blockade on the circadian variation of ventricular arrhythmias. J Am Coll Cardiol 1989; 13:34A.

49. Twidale N, Taylor S, Heddle WF, Ayers BF, Tonkin AM: Morning increase in the time of onset of sustained ventricular tachycardia. Am J Cardiol 1989; 64:1204–1206.

50. Lucente M, Rebuzzi AG, Lanza G, Tamburi S, Cortellessa MC, Coppola E, Iannarelli M, Manzoli U: Circadian variation of ventricular tachycardia in acute myocardial infarction. Am J Cardiol 1988; 62:670–674.

51. Willich SN: Epidemiologic studies demonstrating increased morning incidence of sudden cardiac death. Am J Cardiol 1990; 66:15G–17G.

52. Muller JE, Ludmer PL, Willich SN, Tofler GH, Aylmer G, Klangos I, Stone P: Circadian variation in the frequency of sudden cardiac death. Circulation 1987; 75:131–138.

53. Willich SN, Levy D, Rocco MB, Tofler GH, Stone PH, Muller JE: Circadian variation in the incidence of sudden cardiac death in the Framingham heart study population. Am J Cardiol 1987; 60:801–806.

8

Circadian Rhythm of Sudden Cardiac Death

Robert W. Peters, MD

Epidemiology

Despite major advances in the area of cardiovascular pathophysiology and therapy, sudden cardiac death remains a major public health problem. It has been estimated that almost half a million people die suddenly and unexpectedly in the United States every year.[1] Many of these people are less than 65 years of age, in apparent good health, and might otherwise have been expected to live for many more years. Recent therapeutic developments, such as the use of vasodilators in patients with left ventricular dysfunction, thrombolytic agents in the setting of acute myocardial infarction, and percutaneous transluminal coronary angioplasty (PTCA) and coronary artery bypass surgery in patients with critical coronary artery stenoses, have been demonstrated to prolong life in certain situations. However, these interventions, paradoxically, may actually increase the population at risk of sudden death. One of the difficulties inherent in discussing a topic such as sudden death is that it is not a single entity, but a heterogeneous syndrome, with a multiplicity of causes, only some of which are cardiac in origin (Table 1). All of these etiologies have in common a relatively sudden and unexpected collapse, ultimately culminating in death. Even when restricted to cardiac causes, as will be done in this chapter, the syndrome is far from homogeneous and has many etiologies (Table 2).

From: Deedwania PC (ed): *Circadian Rhythms of Cardiovascular Disorders.*
©Futura Publishing Co., Inc., Armonk, NY, 1997.

Table 1

Noncardiac Causes of Sudden Death

- Cerebrovascular accident
- Acute pulmonary embolus
- Dissecting aortic aneurysm
- Ruptured saccular aneurysm (thoracic, abdominal)
- Obstructed airway (foreign body, asthma)
- Drug overdose
- Anaphylaxis
- Acute massive gastrointestinal bleeding
- Fulminant infection (e.g., meningococcemia)
- Air embolism
- Fat embolism

A major problem in interpreting studies of sudden death is that the premonitory symptoms are often extremely nonspecific, making it difficult to accurately identify a cause in the absence of a detailed postmortem examination (which is rarely available, especially in large epidemiological studies). Thus, the mechanism of sudden death in studies reported in the literature depends on the nature of the population being scrutinized. For example, it is likely that in a clinical trial involving a group of patients with advanced ischemic heart disease in whom an antiarrhythmic drug is to be tested (so that significant conduction system disease is an exclusion criterion), sudden death will most frequently be due to a sustained ventricular arrhythmia. In contrast, the mechanism of sudden death might be quite different in a cohort of young athletes or a group of patients with advanced malignancy. Thus, in the 1970s, when outpatient 24-hour ambulatory electrocardiography (Holter) became routinely available, the mechanism of sudden death was revealed to be ventricular tachycardia or fibrillation in the vast majority of cases. However, since most of the patients undergoing Holter studies had significant cardiac disease, usually arteriosclerotic coronary artery disease, the data obtained in these studies may not be universally applicable.

Another problem in accumulating information about sudden death is that, unlike most other events, the subject is unable to provide historical information. Even in situations where resuscitation is available and successfully carried out (which is certainly the minority of instances), there is often retrograde amnesia. Thus, most of the information that has previously been collected has been obtained through interviews with relatives, bystanders, friends, or review of paramedics' notes or death certificates. This information is often incomplete or se-

Table 2

Causes of Sudden Cardiac Death

Nonarrhythmic Causes
- Ventricular rupture (acute myocardial infarction, trauma)
- Cardiac tamponade
- Acute aortic regurgitation (bacterial endocarditis, trauma)
- Atrial myxoma
- Severe pulmonary hypertension (any cause)
- Aortic stenosis (valvular)
- Hypertrophic subaortic stenosis
- Coronary artery aneurysm (rupture)
- Acute mitral regurgitation (ruptured chordae tendinea, papillary muscle)
- Cyanotic congenital heart disease

Arrhythmic Causes
- Ventricular fibrillation
- Ventricular tachycardia (monomorphic, polymorphic, torsade de pointes)
- Paroxysmal supraventricular tachycardia (associated with preexcitation syndromes, AV nodal reentrant)
- Atrial fibrillation (associated with preexcitation syndromes, accelerated AV nodal conduction)
- Sinus node disease
- Complete AV block

riously flawed, especially in retrospective studies. As will be discussed, some investigators are presently using newer technologies to obtain more definitive information about the chronology of arrhythmic events. The data accumulated in these studies have the potential to provide new information about the mechanisms involved in the pathogenesis of sudden cardiac death.

Pathophysiology

Important insight into the pathophysiology of sudden cardiac death has been provided by recent studies. The atherosclerotic plaque is the hallmark of ischemic heart disease and interest has been focused on the genesis of the plaque and its relationship to the clinical manifestations of ischemic heart disease. The processes involved in plaque formation are well beyond the scope of this discussion, but it is relevant to review some of the available information that pertains to acute cardiac events.

It has recently become apparent that acute cardiac events are precipitated by factors that produce instability in the atherosclerotic plaque, causing extrusion of atheromatous material that may ultimately lead to the formation of an occlusive thrombus. Davies and Thomas performed detailed cardiac postmortem examinations on 100 people who died suddenly, presumably of ischemic heart disease.[2] Coronary artery thrombi were found in 74 and extensive and acute fissuring of atherosclerotic plaques was seen in another 21. Thus, in only 5% of these individuals was there no evidence of acute coronary artery pathology. In contrast, no intraluminal thrombi were found in any of the hearts of age-matched controls who died of other causes. Falk examined the hearts of 47 people with fatal ischemic heart disease and found that the outcome of plaque rupture depended primarily on the degree of preexisting luminal stenosis.[3] When the narrowing was less than 75%, plaque rupture was usually associated with intimal hemorrhage without significant thrombosis but as the stenosis became progressively more severe, an occlusive thrombus was almost invariably present. Falk concludes that severe coronary artery stenoses may serve as a "kind of final common pathway in the atherosclerotic process" leading to the formation of an occlusive thrombus. When examined histologically, these thrombi were found to consist almost entirely of aggregated platelets.

These studies and others in this area help to delineate the relationship between the atherosclerotic process and sudden cardiac death.[4] It is becoming apparent that a variety of activities, including awakening and assuming the upright posture, physical exertion, and the occurrence of emotional distress, which are associated with surges in catecholamine levels (and alterations in vascular reactivity, platelet aggregability, and adhesiveness as well as other interrelated processes) combine to destabilize the atherosclerotic plaque.[5] Extrusion of thrombogenic substances leads to platelet adhesion and aggregation, which, superimposed on the substrate of an already significant luminal stenosis, leads to the formation of an occlusive thrombus. Whether this sudden obstruction produces sudden cardiac death, myocardial infarction, or a clinical syndrome of lesser magnitude depends on the degree of coronary collateralization, the degree of preexisting left ventricular damage, the extent of disease in the remainder of the coronary tree, and probably many other factors as well. The various factors and activities that affect the integrity of the atherosclerotic plaque have been demonstrated to display circadian periodicity.[5] It is hoped that examination of these circadian rhythms will provide further clues to the pathogenesis of these acute events. With this purpose in mind, the growing number of clinical studies of the timing of sudden cardiac death will be discussed in detail.

Clinical Studies (Table 3)

Muller and associates, having described a circadian periodicity in the onset of acute myocardial infarction,[6] were the first to suggest that there may be a similar pattern in the occurrence of sudden cardiac death. They retrospectively analyzed the time of day of the onset of sudden cardiac death (defined as unexpected death occurring within 1 hour of the onset of symptoms) indicated on the death certificates of 2,203 people who died in Massachusetts in 1983.[7] They found that the events that occurred outside of a hospital were distributed in a nonuniform fashion over the 24-hour period with peaks in the mid-morning and in the late afternoon/early evening and a nadir between the hours of midnight and 6 AM. In contrast, deaths in hospitalized patients were relatively evenly distributed throughout the 24-hour period. While there are obvious problems with this type of retrospective data analysis (particularly the possibility that an individual who is found dead in the morning may be listed as having died at the time that his/her death was discovered), the marked similarity to the time of onset of acute myocardial infarction in their previous work suggested that their findings were not an artifact of their method of data collection.

The same group then examined the database from the Framingham Heart Study.[8] Unlike the Massachusetts death certificate study, the Framingham data were derived from a meticulously followed cohort of 5,209 individuals, initially recruited because they lived in Framingham, Massachusetts, and were free of clinically manifest cardiovascular disease.[9] Although Muller's group analyzed the data retrospectively, the mortality information was collected in a carefully standardized manner that included interviews with friends, relatives, and bystanders. They found a circadian rhythm of event onset with a peak in mid-morning and a night-time nadir. This pattern was independent of gender and age and was present both in events labeled as "definite" and those categorized as "possible" sudden cardiac death. Although their findings differed somewhat from their previous study, there is not enough information presented to determine whether this difference could be due to differences in the study populations.

With these retrospective studies as a background, Levine et al. prospectively assessed the timing of cardiac arrest in a consecutive series of patients treated by the city of Houston, Texas, Emergency Medical Service between December 1, 1989, and November 30, 1990.[10] For purposes of analysis, the onset of the event was considered to be the time that the call was first received by the Emergency Medical Service dispatch center. The population included all patients who were found

Table 3

Clinical Studies of the Timing of Onset of Sudden Cardiac Death

1st Author	Subjects (#)	Population	Design
Muller	2203	Mass. death certificates	retrospective
Willich	429	Framingham	retrospective
Levine	1019	EMS in Houston 1 year	prospective
Buff	137	consecutive cpa on med. ward	prospective
Willich	94	death certificate, telephone interview, Mass.	?prospective
Aronow	362	nursing home	prospective
Arnstz	703	consecutive outpatients in Berlin, Germany	?prospective
Couch	1070	autopsies on natives of Kuaui, visitors	retrospective
Gebara	483	ICD database	retrospective
Oeff	515	outpatients treated with external automatic defib.	?
D'Avila	46	ICD database	?
Peters	101	BHAT database	retrospective
Peters	138	CAST database	prospective
Wood	43	ICD database	?

1st Author	AM Peak	PM Peak	BB Adjust	Comments
Muller	7–11	—	no	
Willich	7–9	—	no	careful f/up
Levine	6–12	—	no	
Buff	4–8	8–12	no	
Willich	9–12	—	no	adjusted for awakening time
Aronow	6–12	—	no	elderly population
Arntz	6–12	3–7	no	PM peak only in younger subjects
Couch	6–12	—	no	Kuaui natives
Gebara	9–11	—	no	12,820 episodes, rate unspecified
Oeff	+	+	no	elderly had AM and PM peak, younger had only AM peak
D'Avila	6–12	—	no	
Peters	8–11	—	yes	
Peters	6–12	—	yes	
	—	+		placebo
Wood	—	3	yes	no AA drugs
	—	—		AA drugs

EMS = Emergency Medical Services; AA = antiarrhythmic drugs; f/up = follow-up.

to be in cardiac arrest at the time of arrival of the emergency medical personnel as well as those who developed their arrest after arrival of the personnel or during transportation to the hospital. Specifically excluded were cardiac arrests associated with trauma, toxicological causes (overdoses, poisonings, etc.) or individuals less than 18 years of age. Using harmonic regression analysis, the authors were able to identify a significant mid-morning peak among their 1,019 subjects. Subgroup analysis was apparently not performed.

The above-mentioned studies analyzed primarily those events that began outside of a hospital because of the general perception that hospitalized patients are subjected to a variety of influences that could affect circadian rhythms and also alter the outcome of acute cardiac events. For example, being awakened every 4 hours for vital signs, medications, etc., can markedly affect the sleeping-waking cycle, whereas an episode of ventricular tachycardia, which might be fatal in an outpatient, is often cardioverted without incident in a hospital ward with telemetry capabilities. With this in mind, Buff and associates analyzed the timing of 137 consecutive cardiac arrests that occurred on the medical service of a medium-sized community teaching hospital.[11] Specifically excluded, however, were patients on monitored beds (including those in the emergency department) and patients having major surgery during their hospital stay. They divided the cardiac arrests into those that were sudden and unexpected and those that were expected (patients whose condition was gradually deteriorating in whom death was imminent). The obvious advantage of this type of study is that the timing of events can be determined much more accurately, especially since the data were collected by one of the investigators within 24 hours of the event. The investigators found unexpected cardiac arrests had a peak between 6 AM and 9 AM, whereas the expected cardiac arrests were more evenly distributed throughout the 24-hour period. Muller et al., in contrast, found no significant peaks in events occurring in hospitalized patients but did not divide the patients into expected and unexpected arrests.[7] It is tempting to speculate that the somewhat earlier morning event peak found by Buff and associates (6 AM to 9 AM) compared to some of the other studies (6 AM to 12 noon) may relate to an earlier time of awakening that typically occurs in the hospital (to take vital signs, bathing, etc.). Supporting this hypothesis, Willich and coworkers analyzed the time of onset in 94 individuals experiencing sudden cardiac death in Massachusetts and found that the morning peak in event frequency was actually enhanced by adjusting for the time of awakening.[12] A similar relationship between awakening and event onset has been noted for acute myocardial infarction.[13]

Aronow and Ahn were able to accurately determine the time of unexpected "primary" cardiac arrest (unexpected arrhythmia in a clinically stable patient) in 362 elderly (62–100 years of age) residents of a long-term health care facility, almost 70% of whom were women.[14] In this facility, patients were not on a cardiac monitor but were checked every 30 minutes by nursing personnel. Approximately 40% of events occurred between 6 AM and 12 noon with the remainder distributed evenly among the other three 6-hour periods. Although definitive information is not available, the data from Aronow and Ahn suggest that the morning peak in sudden cardiac death may extend across a variety of subgroups.

Couch describes an 11-year retrospective study of coroners' autopsies in Hawaii in which he compares the temporal distribution of sudden cardiac death (defined as death occurring less than 2 hours after a clinical event) in natives of the island of Kauai to that in visitors.[15] Natives showed a prominent peak between 6 AM and 12 noon, whereas visitors revealed a gradual increase during the day with a peak between 12 noon and 6 PM. The author relates this difference to resetting of the biological clock as well as the stress associated with a prolonged period of travel. The time of awakening on the day of the event was unfortunately unavailable.

Arntz and associates recently reported an interesting study in which they analyzed the time of 703 consecutive sudden cardiac deaths (defined as unexpected collapse unassociated with trauma) attended by emergency medical technicians in the Berlin area between 1988 and 1990.[16] The unique aspect of their study was the routine use of the automatic external defibrillator, a device that records the cardiac rhythm by means of a built-in tape recorder. For purposes of analysis, events were divided into ventricular fibrillation (n=294), asystole (n=260), and pulseless bradyarrhythmias (n=149). They found a bimodal distribution in the ventricular fibrillation group with a significant peak from 6 AM to 12 noon and a secondary peak from 3 PM to 7 PM. The other two groups were more evenly distributed throughout the day although all three groups demonstrated a night-time trough. The bimodal peak in ventricular fibrillation was seen in both genders, but only a morning peak was identified in people less than 65 years of age. The timing differences between the groups have important pathophysiological implications but will require validation by others since the cardiac rhythm that is recorded in unwitnessed arrests is not necessarily the initiating arrhythmia. There was, however, no significant difference between the three groups in the time delay between the alarm and the first recording of the cardiac rhythm.

Further verification of the morning peak in sudden cardiac death is provided by two studies that used third-generation implantable cardioverter defibrillators (ICDs), which provide the exact timing of device activation as well as the cycle length of the initiating arrhythmia. Gebara et al. found a prominent peak in ICD activations between 9 and 11 AM, involving 12,820 episodes in 483 patients over a follow-up period of almost 1 year.[17] Similar findings are reported by D'Avila and coworkers who found that 42% of ICD shocks for very rapid (>180 beats/min) ventricular arrhythmias occurred between 6 AM and 12 noon.[18]

Therapeutic Implications

Inherent in the discussion in this chapter is the hope that furthering our understanding of the epidemiology and pathophysiology of sudden cardiac death will lead to improved preventive strategies. It must also be realized that prevention of sudden cardiac death will necessarily involve preventing myocardial infarction, transient myocardial ischemia, and other entities.

Beta blockers are a widely used class of drugs that have documented efficacy in the prevention of myocardial ischemia and have been demonstrated to reduce mortality following acute myocardial infarction.[19] Because the morning surge in serum catecholamine concentration has been implicated as one of the causative factors in the morning peak in sudden cardiac death, it is logical to try to determine whether beta blockers will eliminate or blunt this peak, as has been shown with acute myocardial infarction.[6] This issue was examined by retrospective analysis of the results of the Beta Blocker Heart Attack Trial (BHAT), a randomized, double-blind, placebo-controlled study designed to determine whether therapeutic doses of the beta blocker propranolol would reduce all-cause mortality when initiated 7–21 days after acute myocardial infarction.[20] The trial was terminated prematurely because a 26% reduction in mortality was found in the active treatment group over a follow-up period of a little more than 2 years. Analysis of the timing of the onset of out-of-hospital sudden cardiac death (Table 4) reveals a clear morning peak in the placebo group that is absent in the patients receiving propranolol. In fact, when the 6-hour time period from 5 AM to 11 AM is eliminated (see asterisks), the number of events in the active treatment and placebo groups is almost identical. Although retrospective, these data suggest that one of the mechanisms by which beta blockers prolong life in patients with ischemic heart disease is by blocking the adverse effects of the morning surge of catecholamines.

The use of antiarrhythmic drugs has also been suggested as a means of reducing the frequency of sudden cardiac death in patients with chronic ischemic heart disease. The Cardiac Arrhythmia Suppression Trial (CAST) was a randomized, double-blind, placebo-controlled study

Table 4

The Beta Blocker Heart Attack Trial (BHAT): Timing of Out-of-Hospital
Sudden Cardiac Death (24-Hour Clock)

Time Interval	Placebo	Propranolol
05:00–08:00*	5	0
08:00–11:00*	15	11
Total	20	11
11:00–14:00	8	5
14:00–17:00	7	7
17:00–20:00	5	3
20:00–23:00	7	7
23:00–02:00	2	8
02:00–05:00	5	3
Total	34	33

*See text for details.

designed to determine whether suppression of ventricular ectopy by means of antiarrhythmic drugs would reduce the incidence of death due to arrhythmias in a high-risk population of patients with ischemic heart disease (left ventricular dysfunction and frequent nonsustained ventricular ectopy following acute myocardial infarction).[21,22] Subjects underwent serial drug testing with encainide, flecainide, or moricizine and those whose ectopy was suppressed ($\geq$80% of ventricular premature beats, $\geq$90% of runs of nonsustained ventricular tachycardia) were randomized to the dose of that drug which suppressed them or to matching placebo and were followed in a double-blind manner. Patients not responding to one drug were tested with the other drugs until a successful regimen was found or until they had failed all doses of all three drugs.

The results of CAST have been reported in detail elsewhere[21,22] and will not be covered in detail in this chapter. Briefly, there was an excess of arrhythmic deaths in the active treatment group which achieved statistical significance after a relatively short follow-up period and which led to premature termination of the trial. The CAST results have triggered a great deal of debate in the cardiology community, most of it focused on the issue of how antiarrhythmic drugs that been demonstrated to have potent antiarrhythmic efficacy in a group of individuals can become (presumably) proarrhythmic in this same group in a relatively short period of time.

Examining the timing of onset of arrhythmic death in the CAST population may help to provide some answers to this problem.[23] When patients receiving beta blockers are excluded, the active treatment group displayed a bimodal distribution (Fig. 1), with a prominent peak

Time of Day: Active Therapy

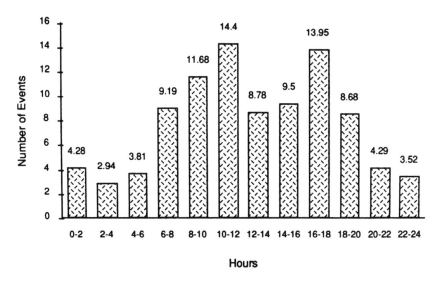

Hours

Figure 1: Timing of event onset (24-hour clock) in the active treatment group. For those events where only a range of time was known, the weight of the event was spread uniformly over the appropriate time. For example, if a patient was known to have died between 7:00 AM and noon, his or her death contributed 0.20 to each of the 5 hours in that interval. (Reprinted with permission.[23])

in mid- to late morning, a secondary peak in the late afternoon and early evening, and a nadir in the late night and early morning hours. This distribution is strikingly similar to that described in previous studies of sudden cardiac death, and especially of acute myocardial infarction, and suggests a possible interaction among sympathetic nervous system activation, the antiarrhythmic drugs used in the study, and possibly acute myocardial ischemia as well.

In contrast, in the placebo group (excluding those receiving beta blockers) the morning peak was conspicuously absent and there was a single broad peak extending from 12 noon to around 8 PM (Fig. 2). That this difference between the active treatment and placebo groups is not artifact is supported by the relationship of event onset to awakening (Figs. 3 and 4). Thus, there is a prominent peak in the first 2 hours after awakening in the active treatment group, whereas the majority of events in the placebo group occurred considerably later.

How do we reconcile the CAST results with previous studies of

Time of Day: Placebo Therapy

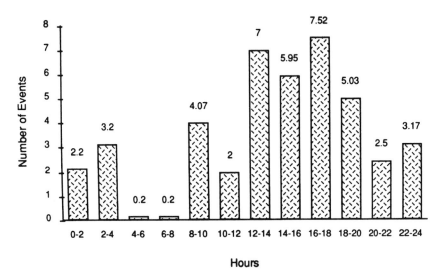

Figure 2: Timing of event onset (24-hour clock) in the placebo group. (Reprinted with permission.[23])

the timing of onset of sudden cardiac death? Unlike most of the other studies, the CAST population was a relatively homogeneous one with the subjects chosen specifically because of their high risk of arrhythmic death. On the other hand, it is likely that the patients who died from arrhythmic causes in CAST were rather similar to those who died suddenly in other studies. The incidence of antiarrhythmic drug use in these other studies is not specified. Could use of these drugs have conceivably affected the results? Would antiarrhythmic drugs not of the 1C class have altered the CAST results? Future studies will be required to answer these important questions. The use of beta blockers in the CAST population was unfortunately not widespread enough to have a significant impact on the timing of arrhythmic death. However, it is noteworthy that arrhythmic deaths in patients receiving beta blockers (although few in number) appear to be scattered throughout the day without any discernible peak.

With the issue of antiarrhythmic drugs in mind, it is worth examining the preliminary data of Wood et al. who describe the timing of 830 activations of third generation ICDs in 43 patients who did not receive beta blockers.[24] In the 22 patients who did not receive antiarrhythmic

Time after awakening:　Active Therapy

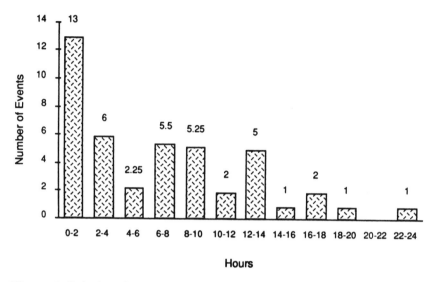

Figure 3: Relation of event onset (24-hour clock) to awakening in the active treatment group. (Reprinted with permission.[23])

drugs, there was a significant afternoon peak (similar to the CAST results?). In the patients who received antiarrhythmic drugs, no peak was discernible. The number of events in each group is not specified but was presumably less in the group that received antiarrhythmic drugs.

Conclusions

What is the message that clinicians can derive from this discussion? Unfortunately, at the present time there are more questions than answers. It is certainly not feasible to avoid awakening, getting up, going to work, and performing the other activities that presumably place people at risk of sudden cardiac death. It has become clear that there is a certain reproducibility to the temporal pattern of the onset of sudden cardiac death but the population to which this pattern applies and the effects of various medications on the pattern need to be further elucidated. Clearly, in the absence of any contraindication, beta blockers should be used prophylactically in patients at high risk of sudden cardiac death, especially in the setting of ischemic heart disease and recent myocardial infarction. Furthermore, although definitive data are not available, it

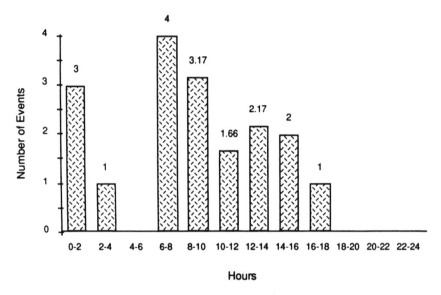

Figure 4: Relation of event onset (24-hour clock) to awakening in the placebo group. (Reprinted with permission.[23])

would seem advisable to employ a medication regimen that would ensure 24-hour coverage, especially since many of the conventional dosing schedules produce the lowest serum concentration at or around the time of the morning awakening when the risk of sudden cardiac death may be at its highest. Thus, it may be that the BHAT results reported above may have underestimated the potential benefits of beta blockers since the medication in that study was administered on a t.i.d. schedule.

Further investigation is clearly warranted in certain areas. For example, the effect of other interventions that may improve survival in certain types of patients at high risk of sudden cardiac death (e.g., vasodilators in congestive heart failure, coronary artery angioplasty, coronary artery bypass surgery) on the timing of sudden cardiac death needs to be delineated. The effect of the day of the week, the season of the year, and other factors that have been demonstrated to affect the onset of acute myocardial infarction[25] upon sudden cardiac death is another fertile area for investigation. It will also be important to determine whether the patterns of onset of sudden cardiac death discussed above are universally applicable or apply only to certain segments of the population. Equally crucial will be differentiating between the patterns of onset of sudden cardiac death due to bradyarrhythmias from that due to ventricular tachycardia or fibrillation. Future studies in these and other important areas will hopefully add to the accumulating database on the chronology of sudden cardiac death and eventually be translated into therapeutic strategies.

References

1. American Heart Association: 1992 Heart and Stroke Facts. Dallas: The American Heart Association, 1992.
2. Davies MJ, Thomas A: Thrombosis and acute coronary-artery lesions in sudden ischemic death. N Engl J Med 1984; 310:1137.
3. Falk E: Plaque rupture with severe pre-existing stenosis precipitating coronary thrombosis: characteristics of coronary atherosclerotic plaques underlying fatal occlusive thrombi. Br Heart J 1983; 50:127.
4. Muller JE, Abela GS, Nesto RW, Tofler GH: Triggers, acute risk factors and vulnerable plaques: the lexicon of a new frontier. J Am Coll Cardiol 1994; 23:809.
5. Muller JE, Tofler GH, Stone PH: Circadian variation and triggers of onset of acute cardiovascular disease. Circulation 1989; 79:733.
6. Muller JE, Stone PH, Turi ZG, et al: Circadian variation in the frequency of onset of acute myocardial infarction. N Engl J Med 1985; 313:1315.
7. Muller JE, Ludmer PL, Willich SN, et al: Circadian variation in the frequency of sudden cardiac death. Circulation 1987; 75:131.
8. Willich SN, Levy D, Rocco MR, et al: Circadian variation in the incidence of sudden cardiac death in the Framingham Heart Study population. Am J Cardiol 1987; 60:801.
9. Dawber TR, Meadors GF, Moore FE Jr: Epidemiologic approaches to heart disease: the Framingham study. Am J Pub Health 1951; 41:279.
10. Levine RL, Pepe PE, Fromm RE Jr, et al: Prospective evidence of a circadian rhythm for out-of-hospital cardiac arrests. J Am Med Assn 1992; 267:2935.
11. Buff DD, Fleisher JM, Roca JA, et al: Circadian distribution of in-hospital cardiopulmonary arrests on the general medical ward. Arch Intern Med 1992; 152:1282.
12. Willich SN, Goldberg RJ, Maclure M, et al: Increased onset of sudden cardiac death in the first 3 hours after awakening. Am J Cardiol 1992; 70:65.
13. Peters RW, Zoble RG, Liebson PR, et al: Identification of a secondary peak in myocardial infarction onset 11 to 12 hours after awakening: the Cardiac Arrhythmia Suppression Trial (CAST) Experience. J Am Coll Cardiol 1993; 22:998.
14. Aronow WS, Ahn C: Circadian variation of primary cardiac arrest or sudden cardiac death in patients aged 62 to 100 years (mean 82). Am J Cardiol 1993; 71:1457.
15. Couch RD: Travel, time zones, and sudden cardiac death: emporiatric pathology. Am J Forensic Med Pathol 1990; 11:106.
16. Arntz HR, Willich SN, Oeff M, et al: Circadian variation of sudden cardiac death reflects age-related variability in ventricular fibrillation. Circulation 1993; 88:2284.

17. Gebara OCE, Mittelman M, Rasmussen C, et al: Morning peak in ventricular arrhythmias detected by time of implantable cardioverter-defibrillator therapy. J Am Coll Cardiol 1994; 23:204A.
18. D'Avila A, Gebara O, Brugada P: Circadian variation in recurrent sudden death aborted by discharges from the implantable defibrillator. Circulation 1993; 88:I-155.
19. Beta-Blocker Heart Attack Trial Research Group: A randomized trial of propranolol in patients with acute myocardial infarction: mortality results. J Am Med Assoc 1982; 247:1707.
20. Peters RW: Propranolol and the morning increase in sudden cardiac death: (the Beta Blocker Heart Attack Trial Experience). Am J Cardiol 1990; 66:57G.
21. The CAST Investigators: Preliminary report: effect of encainide and flecainide on mortality in a randomized trial of arrhythmia suppression after myocardial infarction. N Engl J Med 1989; 321:406.
22. The CAST Investigators: Effect of the antiarrhythmic agent moricizine on survival after myocardial infarction. N Engl J Med 1992; 327:227.
23. Peters RW, Mitchell LB, Brooks MM, et al: Circadian pattern of arrhythmic death in patients receiving encainide, flecainide or moricizine in the Cardiac Arrhythmia Suppression Trial (CAST). J Am Coll Cardiol 1994; 23:283.
24. Wood MA, Simpson PM, London WB, et al: Circadian pattern of ventricular tachyarrhythmias in patients with implantable cardioverter defibrillators. J Am Coll Cardiol 1994; 23:204A.
25. Peters RW, Zoble RG, Brooks MM, et al: Increased onset of acute myocardial infarction on Monday. Circulation 1993; 88:I-509.

9

Circadian Variation in Stroke Onset

John Marler, MD

Introduction

Stroke is a clinical syndrome characterized by the sudden onset of a focal neurological deficit. This syndrome can be caused by several different pathological mechanisms. The most common is the occlusion of a brain artery by a blood clot that either forms in situ or is carried from a more proximal site in the vascular tree to lodge in a brain vessel. This is called ischemic stroke because the brain is injured by a lack of blood. Another common cause of stroke is hemorrhage from an artery or arteriole within the brain parenchyma. This is called intracerebral hemorrhage. Stroke can also be caused by the rupture of an aneurysm in a brain artery in the subarachnoid space adjacent to the brain. In addition to these three most common causes of the stroke syndrome, there are many other less common causes.

The time of onset is more easily determined for stroke than for other disease processes that have a more gradual onset. In just a few minutes, serious neurological impairments can develop. Although there is some resolution in a large number of cases, the onset time can usually be estimated after interviewing either the patient or the family members.

Clinical researchers have recorded the onset time of stroke and analyzed for the presence of a time when strokes are more likely to occur.

From: Deedwania PC (ed): *Circadian Rhythms of Cardiovascular Disorders.* ©Futura Publishing Co., Inc., Armonk, NY, 1997.

The reasons for doing so have been to gain some further understanding of the mechanisms causing stroke and to make plans for routine and emergency treatment of stroke.

Problems

There are several difficulties encountered in determining the time of onset of stroke and whether there is an associated circadian variation. Some of the problems are practical and some are conceptual.

The foremost practical problem is the dependence on self-reporting. For one reason or another, there may be a tendency to underreport strokes that occur during a certain time of the day. There may be a greater likelihood that an individual with a stroke will go to a hospital at one time of the day than at another time of day. Patients may more accurately recall strokes with more sudden onset than those with onset that is more gradual. An accurate distinction may not be made between strokes that are present on awakening and those that occur early in the day. It is not known how accurate a report of awakening without a deficit actually is. A patient with onset of aphasia and a mild paresis of the right arm may report being normal until sitting down to breakfast and attempting to talk or eat. Previously existing symptoms may be perceived as having the same time of onset as the time of realization that the symptoms were present. If the time of stroke onset is to have any interpretation as a consequence of a biological mechanism, then it is important that the observations reported accurately reflect changes in the biological system rather than the acuity of the perception or the receptiveness of the medical care delivery and reporting. To a certain extent, these problems can be addressed by doing many different studies and looking for similar results. Bias in reporting can be limited to some extent in doing prospective studies by obtaining complete data on a large series of consecutive patients or by doing population-based rather than hospital-based studies.

Conceptual problems are encountered with the strokes that might occur during sleep and largely resolve with minimal or ill-defined symptoms by the time the patient is able to report any symptoms later the next morning. There is also the possibility that there is delay of several hours or more between the stroke-precipitating event and the secondary events that cause symptoms. For example, a factor that has a circadian variation in activity may have peak activity at 1400 in the afternoon. If a stroke caused by the 1400 event does not become symptomatic until 0900 the next morning, then effort could be wasted trying to identify inciting factors that have peak or trough activity at 0900 rather than at 1400. The connection time between an inciting factor and a subsequent stroke must remain very speculative and, for the time being, unknown. It is generally assumed that the time from an extreme inciting factor activity to the onset of stroke is very short and does not involve the passage of hours.

Clinical Observations

Numerous studies have documented the time of day that stroke symptoms are first noted. Various designs have been used to address the problems that are inherent in this type of clinical research. Results of some significant studies are described in the following section.

The report by Wroe, Sandercock, et al.[1] is an important recent addition to our knowledge of the time of onset of all three common types of stroke. This study has the advantage that it is a community-based study rather than a hospital-based study. Stroke was observed in 675 individuals over 4 years. Ischemic stroke occurred in 545, intracerebral hemorrhage in 66, subarachnoid hemorrhage in 33, and the type of stroke was unknown in 31. In 554 patients (82.1%), the time of onset was determined. For 578 patients (85.6%), it was known whether onset occurred while asleep or awake. The proportion with onset during sleep was 25% (135/545) for ischemic stroke, 17% (11/66) for intracerebral hemorrhage, and 0% (0/33) for subarachnoid hemorrhage. For all stroke types together there was a significant ($chi^2 = 218.7$, p <0.001) diurnal variation with a morning peak between 0800 and 1000. Significant diurnal variation was also found in the onset of ischemic stroke (peak 0800–1000, $chi^2 = 208.4$, p <0.001). Fewer patients had other forms of stroke and the diurnal variations for primary intracerebral hemorrhage (peak 1000–1200) and subarachnoid hemorrhage (peaks 0800–1000 and 1800–2000) were not significant. A possible secondary peak in stroke onset time was observed for ischemic stroke between 1800 and 2000.

In a report from India, Pardiwalla[2] describes stroke onset times for 148 ischemic and 34 hemorrhagic strokes seen within 12 hours of onset. For both stroke types the frequency of onset was found to be highest between 0601 and 1400. Strokes present on awakening were counted as occurring from 2201 to 0600.

A prospective study of stroke onset time in 977 cases of acute stroke seen in an emergency room was reported from Italy by Gallerani et al.[3] Stroke onset was categorized to the nearest hour, if possible, otherwise to the nearest 6-hour period defined by midnight, 0600, noon, and 1800. Diagnosis was confirmed by CT scan. Stroke etiology was determined clinically and categorized as ischemic stroke, hemorrhagic stroke, or subarachnoid hemorrhage. This reports the use of the cosine method for analysis of time-qualified frequency series. Essentially, a cosine curve is fitted to the 24-hour data by least sum of squares methodology. The report cited contains a more detailed explanation of the methodology. The report indicates that 38.5% of strokes occurred between 0700 and 1200. When analyzed separately, cerebral hemorrhage did not show a significant circadian rhythm. This report confirms previous observations in a large population and in a prospective study. It demonstrates a method for statistical analysis not used in other similar studies.

Another Italian study reported by Ricci et al.[4] was community based. The authors were seeking a more accurate estimate of time of onset without the bias introduced by limiting observations to hospitalized patients. In a population of approximately 49,000, adequate data about the time of onset was available for 368 individuals during the years 1986–1989. For 281 patients with ischemic stroke, 55 (19.6%) had stroke onset while asleep. The most frequent 3-hour interval of stroke onset was 0600 to 0900. The second most frequent 3-hour interval of stroke onset was 0900 to 1200. Intracerebral hemorrhage and subarachnoid hemorrhage also tended to occur in the morning. The authors concluded that referral bias did not change the apparent morning peak in stroke onset time.

A retrospective hospital study was reported by van der Windt from the Netherlands.[5] For 66 patients, stroke onset occurred between 0600 and 1800 in 46 (78%) patients. The number of strokes in the period 0600–1200 was 22 while the number in the period 1200–1800 was 24, which is almost the same.

A retrospective case series of ischemic stroke onset time in patients with cardiac valve replacement was reported by Butchart et al. from Wales.[6] Of 622 patients who survived mitral valve replacement, 96 suffered 139 ischemic cerebral events including stroke and transient ischemic attack (TIA). Time of onset was able to be determined for 108 events in 67 patients. Time of onset was classified into 6-hour categories. The morning 0600 to 1200 period showed the most frequent onset of ischemic cerebral events.

Prospective data on the time of onset and whether or not onset was during sleep were collected from 1,805 hospitalized stroke patients soon after the onset of symptoms. Of these were 1,273 patients with ischemic stroke, 196 with subarachnoid, and 237 with intracerebral hemorrhage. The 2-hour period with most frequent time of onset for ischemic stroke was between 0800 and 1000.[7] For stroke caused by subarachnoid and intracerebral hemorrhage, the most frequent onset time was between 1000 and 1200.[8]

Marsh et al.[9] reported prospective observations on 151 hospital patients seen within 24 hours of ischemic stroke onset. The most frequent time of onset for all ischemic stroke subtypes was the period from 0600 to 1200. They observed that the subgroup of 46 patients taking aspirin had the same apparent peak time of onset. By seeing the patients so soon after hospitalization, any bias due to loss of information with the passage of time may have been diminished.

In a letter, Ince[10] reported a study of 110 patients. This letter reported that more strokes occurred from 0600 to 1800 than from 1800 to 0600. No increased incidence was noted in morning hours. The author suggested that the results had increased reliability because most older individuals lived with their families and because most observed strict Islamic rules, which required prayer at five specific times throughout the day. This allowed for more frequent and reliable observation prior to stroke than is available in other cultures.

Two early studies reported periods of maximum stroke onset that are different than those reported in most other studies. Hossmann[11] reported a study based on a register of 131 hospitalized patients with ischemic stroke. Peak incidence was between 0100 and 0500 during the hours when most patients were asleep. He hypothesized that stroke was caused by hypoperfusion due to the decline in blood pressure that occurs during sleep. Marshall[12] reported results on a retrospective study of 707 patients that had most frequent onset of stroke in the period from 0000 to 0600. There are other early reports that showed onset to occur most often later in the morning after 0600.[13–16]

The conclusions of these clinical reports are summarized in Figure 1. The computed average reflects the average period of maximum onset that was reported. It does not include any weighting to reflect the number of patients in each study. The two reports that show a very early onset time were included in the calculation of the average time, which is approximately at 0900. The intervals used in Table 1 and Figure 1 are those chosen by the authors. In some cases, graphs in the reports showed a peak frequency of onset in a smaller interval, but this was always included within the larger interval that was reported.

In summary, there is an increasing amount of evidence from many studies that stroke occurs most often in the late morning between 0600 and 1200.

Reported Periods of Highest Stroke Frequency

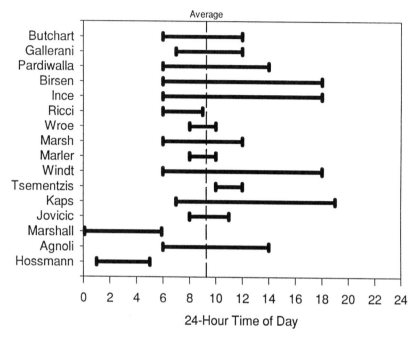

Figure 1: Observed range of most frequent stroke onset as reported in respective studies (see Table 1).

Table 1

Peak Onset Time of Stroke

Author/Reference #	Year	Sample Size	Peak Onset Time*
Hossmann (11)	1971	131	01:00–05:00
Agnoli et al. (13)	1975	256	06:00–14:00
Marshall (12)	1977	707	00:00–06:00
Jovicic (14)	1983	85	08:00–11:00
Kaps (16)	1983	545	07:00–19:00
Tsementzis (15)	1985	245	10:00–12:00
van der Windt (5)	1988	66	06:00–18:00
Marler (7)	1989	1167	08:00–10:00
Marsh (9)	1990	151	06:00–12:00
Wroe (1)	1992	554	08:00–10:00
Ricci (4)	1992	368	06:00–09:00
Ince (10)	1992	110	06:00–18:00
Birsen (10)	1992	110	06:00–18:00
Pardiwalla (2)	1993	182	06:00–14:00
Gallerani (3)	1993	977	07:00–12:00
Butchart (6)	1994	108	06:00–12:00

*All times are based on 24-hour clock.

Inciting Factors

Epidemiological studies have proven that several factors may increase an individual's risk for stroke. These factors include hypertension, smoking, diabetes, and atrial fibrillation. However, knowledge of these factors does not allow conclusions to be made about the actual mechanism, the sequence of events, that causes a stroke. Knowledge of the first step in the process, the inciting factor, would add immeasurably to our knowledge of stroke and would open many avenues for the development of new preventive therapies. The demonstration of a circadian variation in the time of stroke onset could indicate a specific inciting event that occurs with increased frequency at a predictable time.

Not all individuals with risk factors such as hypertension, atrial fibrillation, tobacco use, or diabetes have strokes. They certainly do not all have a stroke every day between 6 AM and noon. Some additional factor, or more likely a constellation of factors, must occur to initiate a process that leads to stroke in a person who has been at risk but stroke-free for years. This factor or factors must stay within stroke-free boundaries except on rare occasions. During these rare occasions, an unusual combination of events creates a situation that takes an individual beyond the capability of normal homeostatic mechanisms to protect and maintain brain circulation.

These factors involved will be discovered one at a time. The presence of a circadian variation for the time of stroke onset may be interpreted as a clue that some of the events that precipitate the occurrence of a stroke occur in a circadian pattern. For instance, it has been suggested that the level of fibrinogen is a risk factor for stroke.[17–19] If a well-tolerated treatment were found to reduce the daily variation in fibrinogen levels, it might be considered for further pilot testing to see whether the new treatment actually showed some promise as a method to prevent stroke. Careful preliminary studies would have to be done in order to justify this conclusion since one possible result is that strokes which usually occurred in the morning would now occur throughout the day. Another more likely result would be that however appealing and logical, reducing the daily variation of a particular factor may have no effect on the likelihood of a stroke.

Daily variations in blood pressure are cited as a possible factor in the circadian variation in the time of stroke onset. In one report, the acrophase for systolic blood pressure in normal adults was 18:24 ± 00:36 (SEM).[20] Shimada et al.[22] reported on blood pressure changes in 88 elderly patients. Of these, 54 were hypertensive. Hypertensive patients were classified as "dippers" if the mean awake systolic pressure was 10 mm Hg greater than the mean while asleep. There were 15 hypertensive "nondippers" and 39 hypertensive "dippers." MRI scans showed 3.7 lacunar infarctions per patient in the "nondipper" group versus 1.0 and 0.9 per patient in the hypertensive "dippers" group and all normotensive patients. The "nondipper" patients had significantly elevated average 24-hour blood pressures compared to the other two groups; however, the average awake blood pressure was similar in the two hypertensive "dipper" and "nondipper" groups. The conclusion of the report was that in hypertensive patients, the sleeping, rather than the waking, blood pressure may be a more important variable. If these observations are valid, the challenge is to understand the relationship of a condition early in the morning to the onset of symptoms several hours later.

Table 2 lists many of the inciting factors that have been proposed to explain the late morning peak in stroke onset. Designing research protocols to scientifically address these hypotheses will be difficult, but if answers are found, the benefit to those at risk of stroke could be very great.

Future Research

There are several avenues for exploration in the future. The final goal of any clinical research would be to intervene to prevent or treat disease, in this case, stroke. Further observation of clinical stroke without intervention can proceed in several directions. Further studies done only to further document and characterize the circadian pattern of

---

Table 2

Possible Causes of Diurnal Variation Onset Triggering Mechanisms

Circadian changes in blood pressure
Decreased blood pressure during sleep
Morning rise in blood pressure
Packed cell volume
Plasma fibrinolytic activity
Platelet aggregability
Morning activity
Moderate to heavy exercise
Increased activity of the clotting system
Plasma catecholamine activity
High morning plasminogen fast acting inhibitor
Low morning TPA activity
Viscoelasticity of whole blood
Increased sympathetic tone
Decreased antithrombin III
Central dopamine activity

stroke onset will be of less interest. If such descriptive studies are to add further information to the field, they should probably be larger population-based studies designed to eliminate any observer and reporting biases that might have skewed the results that have been reported so far.

A more productive direction for future research might be to explore in more detail the correlation between biochemical and physiological circadian rhythms and the pattern observed for the time of onset of stroke. Such studies could actually help define precipitating events that actually lead individuals with risk factors such as hypertension or atrial fibrillation to have a stroke at a particular time. As our knowledge of the disease processes that cause stroke becomes more complete, complex relationships will be observed.

Knowledge about the circadian variation in the time of onset is already being used to develop an intervention for myocardial infarction.[21] Treatments are being directed at either diminishing or removing triggering factors and at protecting individuals during the time of highest risk.

The most useful results of the circadian variation would be the development of knowledge of new risk factors that would predict those most at risk of stroke and the development of new understanding of the mechanism of stroke that would lead to new interventions to prevent stroke in the first place.

References

1. Wroe SJ, Sandercock P, Bamford J, et al: Diurnal variation in incidence of stroke: Oxfordshire community stroke project. Br Med J 1992; 304(6820): 155–157.
2. Pardiwalla FK, Yeolekar ME, Bakshi SK: Circadian rhythm in acute stroke. J Assoc Physicians India 1993; 41(4):203–204.
3. Gallerani M, Manfredini R, Ricci L, et al: Chronobiological aspects of acute cerebrovascular diseases. Acta Neurol Scand 1993; 87(6):482–487.
4. Ricci S, Celani MG, Vitali R, et al: Diurnal and seasonal variations in the occurrence of stroke: a community-based study. Neuroepidemiology 1992; 11(2):59–64.
5. van der Windt C, van Gijn J: Cerebral infarction does not occur typically at night. J Neurol Neurosurg Psychiatry 1988; 51:109–111.
6. Butchart EG, Moreno de la Santa P, Rooney SJ, et al: The role of risk factors and trigger factors in cerebrovascular events. J Cardiac Surg 1994; 9(2 Suppl):228–236.
7. Marler JR, Price TR, Clark GL, et al: Morning increase in onset of ischemic stroke. Stroke 1989; 20(4):473–476.
8. Sloan MA, Price TR, Foulkes MA, et al: Circadian rhythmicity of stroke onset: intracerebral and subarachnoid hemorrhage. Stroke 1992; 23(10): 1420–1426.

9. Marsh EE III, Biller J, Adams HP, et al: Circadian variation in onset of acute ischemic stroke. Arch Neurol 1990; 47:1178–1180.
10. Ince B: Circadian variation in stroke [letter; comment]. Arch Neurol 1992; 49(9):900.
11. Hossmann V: Circadian changes of blood pressure and stroke. In: Zulch KJ (ed), Cerebral Circulation and Stroke. Berlin Heidelberg/New York, Springer, 1971, pp. 203–208.
12. Marshall J: Diurnal variation in occurrence of strokes. Stroke 1977; 8:230–231.
13. Agnoli A, Manfredi M, Mossuto L, et al: Rapport entre les rhythmes homoronyctaux de la tension arterielle et sa pathogonie de l'insuffisance vasculaire cerebrale. Rev Neurol (Paris) 1975; 13(1):597–606.
14. Jovicic A: Bioritam i ischemicni crebrovaskularni poremcaji. Vojnosanit Pregled 1983; 40:347–351.
15. Tsementzis SA, Gill JS, Hitchcock ER, et al: Diurnal variation of and activity during the onset of stroke. Neurosurgery 1985; 17:901–907.
16. Kaps M, Busse O, Holmann O: Zur circadianen haufigkeitsverteilung ischamischer insulte. Nervenarzt 1983; 54:655–657.
17. Eber B, Schumacher M: Fibrinogen: its role in the hemostatic regulation in atherosclerosis. Semin Thromb Hemost 1993; 19(2):104–107.
18. Wilhelmsen L, Svärdsudd K, Korsan-Bengtsen K, et al: Fibrinogen as a risk factor for stroke and myocardial infarction. N Engl J Med 1984; 311(8):501–505.
19. Kannel WB, Wolf PA, Castelli WP, et al: Fibrinogen and risk of cardiovascular disease: The Framingham Study. JAMA 1987; 258:1183–1186.
20. Cugini P, Di Palma L, Di Simone S, et al: Circadian rhythm of cardiac output, peripheral vascular resistance, and related variables by a beat-to-beat monitoring. Chronobiol Int 1993; 10(1):73–78.
21. Muller JE, Tofler GH: Triggering and hourly variation of onset of arterial thrombosis. Ann Epidemiol 1992; 2(4):393–405.
22. Shimada K, Kawamoto A, Matsubayashi K, et al: Diurnal blood pressure variations and silent cerebrovascular damage in elderly patients with hypertension. J Hypertension 10(8):875–878, 1992.

10

Chronobiology and Chronotherapeutics:

Applications to Cardiovascular Medicine

Michael H. Smolensky, PhD

Introduction

The concept of homeostasis postulates that there is constancy of the *intern milieu*. It is a powerful construct in biology influencing the teaching and understanding of the medical sciences as well as the practice of clinical medicine. Thus, it is assumed the risk and exacerbation of disease are invariable and independent of the time of day, day of month, and month of year as are the responses of patients to diagnostic tests and medications. Findings from the field of chronobiology, the study of biological rhythms, challenge the concept of homeostasis and the many assumptions and procedures of clinical medicine based on it. It is now recognized that human functions have daily, weekly, monthly, and yearly biological rhythms (Table 1). Plants, animals, and insects also have chronobiological rhythms.[1-3]

The majority of physicians, nurses, and pharmacists are unfamiliar with the field of chronobiology. This is because the teaching of biology in schools of medicine, nursing, and pharmacy is based solely on the theory

From: Deedwania PC (ed): *Circadian Rhythms of Cardiovascular Disorders*. ©Futura Publishing Co., Inc., Armonk, NY, 1997.

Table 1

Illustrative Spectrum of Biological Rhythms*

Domain	High Frequency	Medial Frequency	Low Frequency
Major Rhythmic Components:	$\tau < 0.5h$ $\tau \sim 0.1s$ $\tau \sim 1s$ et cetera	$0.5h < \tau < 6d$ Ultradian (0.5 $< \tau <$ 20h) Circadian (20 $< \tau <$ 28h) Infradian (28 $< \tau <$ 6d)	$\tau > 6d$ Circaseptan ($\tau \sim$ 7d) Circamensual ($\tau \sim$ 30d) Circannual ($\tau \sim$ 1yr)
Examples of Rhythms In:	Electroencephalogram Electrocardiogram Respiration	Rest-activity Sleep-wakefulness Responses to drugs Blood constituents Urinary variables Metabolic processes, generally	Menstruation 17-Ketosteroid excretion with spectral components in all regions indicated above and in other domains

Spectrum of human biological rhythms. High frequency rhythms (Greek tau symbolizes period and mathematically equals 1/frequency) in the range of a second or less are exemplified in the tracings of the electrocardiogram and electroencephalogram. Medial frequencies include ultradian rhythms, exemplified by pulsatile hormonal secretions and progression of sleep stages, and circadian rhythms, which are common to nearly all living functions. Low frequency rhythms include 7-day (circaseptan), monthly (circamensual), and 1-year (circannual) rhythms; many endocrine and immune functions exhibit one or more of these rhythms. Reproduced from Halberg[4] with permission.

of homeostasis. We are taught that our biology is maintained in a relatively constant state by inherited mechanisms that are activated when internal conditions deviate from specific biological set points. Chronobiology, on the other hand, teaches that human biological functions and processes exhibit predictable-in-time, cyclic variability. The conceptual differences between homeostasis and chronobiology are not merely of academic interest. Emerging findings from chronobiological investigations are of such importance that reexamination of several fundamental assumptions and practices of clinical medicine is warranted.

Human Chronobiology: Concepts and Principles

Before addressing the relevance of chronobiology to medicine, particularly cardiovascular disease, it is necessary to define the key terms and concepts that are central to this new science.[4-6]

A biological rhythm is a self-sustaining oscillation with the period, that is, the duration of time between successive repetitions, being rather nonvarying under normal conditions. Biological rhythms are defined by a specific set of characteristics (Fig. 1). The first is *period*—24 hours or 28 days, for example. One of the most important bioperiodicities in medicine is the circadian (circa = about; dies = day, or an approximate 24-hour) rhythm. Almost every endpoint used in

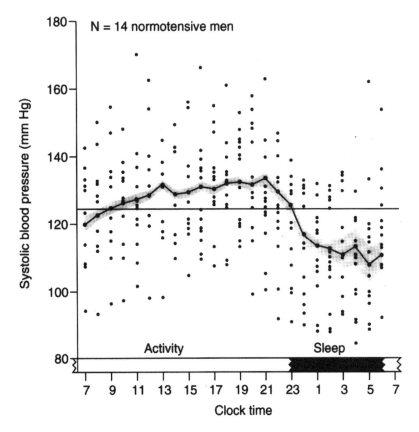

Figure 1: Descriptive parameters of the circadian rhythm of systolic blood pressure (SBP) exemplified by the data of 14 diurnally active (0629–2309) normotensive persons assessed by 24-hour ambulatory monitoring. Hourly mean SBPs are shown as small dark circles. The continuous sine-wave curve connects the hourly clock-time group means and the shading indicates X ±1 SE. The solid horizontal straight line depicts the group 24-hour mean (124 mm Hg) around which rhythmicity occurs. In this group, blood pressure was greatest at 2100 (134 mm Hg) and lowest at 0500 (108 mm Hg). The peak-to-trough difference, or amplitude, amounted to 26 mm Hg. (Reproduced with permission from Thomas et al.[7])

the diagnosis of human disease and assessment of treatment outcomes varies in a systemic manner during the 24 hours. A second characteristic is *amplitude*—the predictable variability over time ascribable to rhythmicity. Some circadian rhythms, such as those in body temperature and heart rate, are of relatively low amplitude; others, such as those of plasma cortisol, adrenaline and lymphocyte number, are of quite high amplitude. A third characteristic is *level* (baseline) around which the predictable variability in time is manifested. Finally, a rhythm is characterized by its *phasing,* that is, the occurrence of the peak and/or trough values with reference to a given time scale—24 hours or a month (menstrual cycle), for example. Collectively, chronobiologists refer to the intricate rhythmic organization of life's functions and processes as the biological time structure.

Initially, many classically trained biologists believed that rhythms were conditioned responses of the body to cyclic phenomena presented by the ambient environment.[7,8] This is not the case. Life forms inherit by genetic transmission specific clock mechanisms that drive observable biological rhythms.[9–11] The period and phasing of rhythms are coordinated by pacemaker clocks located at various levels of biological organization, with those in the brain being dominant.[12,13] Human circadian clocks are set or reset on a day-to-day basis by environmental time cues termed "synchronizers," the most important being the daily timing of lights on (sunrise) and off (sunset) in conjunction with one's wake-sleep routine.[14,15] Circadian rhythms of persons dwelling under experimental conditions devoid of time cues, such as found in caves or special study bunkers for example, continue. However, in the absence of normal social and environmental time cues, circadian clocks tend to free-run, giving rise to periodicities slightly longer or shorter than 24 hours, which is the norm in the usual environment.[14–16] Under constant environmental conditions, the phase-relationships between all of the body's circadian processes and functions become scrambled and desynchronized from normal.[16,17] This same type of phenomenon, desynchronization of the circadian time structure, is experienced temporarily by workers and travelers who suddenly experience alteration of their activity-sleep cycle due to shift-work rotation[18] or rapid displacement across several time zones by high-speed aircraft; the latter is termed "jet lag."[19] The observations that biological rhythms persist in human and other organisms, even when the environment is devoid of time cues, as well as undergo change in phase with alteration in the sleep-wake routine, constitute evidence that rhythms are endogenous in origin rather than conditioned responses to environmental cycles of like period.

Although the biological time structure is genetically determined, its expression is determined by several factors. For example, the inherited amplitude of the endogenous circadian rhythm of blood pressure in most healthy persons is of relatively small magnitude. Day-night differences in activity and stress amplify the innate biological variability; in normal persons the 24-hour, peak-to-trough variation in systolic and diastolic pressures commonly amounts to 15–25 mm Hg.[20–23] Disease also affects the expression and characteristics of circadian rhythms. In hypertensive patients, both the 24-hour mean level and amplitude of the rhythm may be altered; in some cases the rhythm itself might be obliterated.[23,24] The 24-hour blood pressure pattern of essential hypertension is characterized by a peak during daytime activity and a dip (trough) during night-time sleep. In secondary hypertension, the magnitude of the sleep-time dip is often blunted or obliterated; indeed, there may even be an elevation in blood pressure over daytime values resulting in an altered phasing of the circadian rhythm, as shown in Figure 2.[25]

The phasing of human circadian clocks and rhythms is set, or said to be synchronized, primarily by the sleep in darkness-activity in light, 24-hour routine. Shift workers, when assigned to night duty, adhere to a different sleep-activity routine than when assigned to days. The occurrence of the peak and trough of circadian rhythms, with reference to external clock-time, changes in conjunction with alteration of the sleep-wake routine.[18] In personnel working the morning shift from 0600 to 1400, the peak of the rhythm in systolic and diastolic blood pressure occurs early in the afternoon, about 1400. When working nights from 2200 to 0600, the peak occurs at a different time of day, 0600 (Fig. 3). Although the clock-hour occurrence of the circadian peak in the blood pressure rhythms differs when workers are on the day and night shifts, it is found to be comparably timed when referenced to the sleep-wake routine of the respective work shifts. The peak occurs approximately 8 hours following awakening from sleep, whether it is taken during the night or during the day.[20] Thus, the sleep-wake cycle is a useful marker or reference for estimating the staging of rhythms comprising the human circadian time structure. This point is important for the proper application of chronobiology in medicine, for example, the interpretation of diagnostic test results that might be influenced by high-amplitude circadian rhythms or when prescribing medications to meet circadian rhythm determinants.

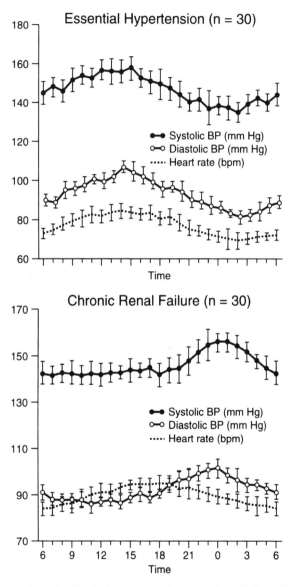

Figure 2: Circadian rhythm in heart rate plus systolic and diastolic blood pressure (X ±1 SD) in 30 diurnally active essential (top) and 30 renal hypertensives (bottom). Essential hypertensives exhibited rhythms with afternoon peaks and nocturnal (sleep time) troughs. The blood pressure rhythms were differently phased in patients with hypertension secondary to renal disease. They peaked during night-time sleep with lowest values during daytime activity. The heart rate circadian rhythm, with afternoon peak, was preserved however. (Reproduced with permission from Portaluppi et al.[25])

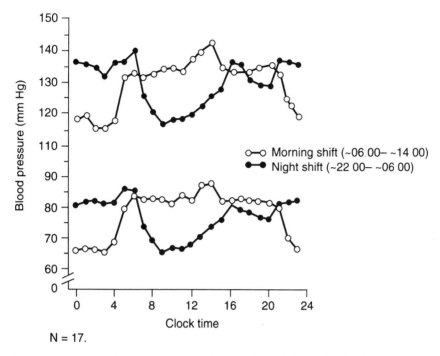

N = 17.

Figure 3: Twenty-four-hour pattern in systolic and diastolic blood pressure in 17 rotating shift personnel assessed while working morning from ~ 0600 to ~ 1400 (open circles) and night from ~2200 to ~0600 (closed circles) work schedules. On the morning shift, blood pressure peaked ~1400, at the end of the work period; it was lowest during night-time sleep (~0200). On the night shift, blood pressure peaked at a different clock time, ~0600, but nonetheless, at the end of work. It was lowest during the daytime (~0900). (Reproduced with permission from Baumgart.[20])

Circadian Rhythms in the Occurrence and Severity of Disease

The physiology and biochemistry of human beings vary greatly during a 24-hour cycle. Circadian rhythms in critical bioprocesses give rise to significant day-night patterns in the manifestation and severity of many common human diseases and their symptoms (Fig. 4). Among these are:

- Allergic rhinitis, with the major symptoms of sneezing, runny nose, and stuffy nose being at least twice as great in intensity during the morning when arising from night-time sleep than during the afternoon and evening;[26]

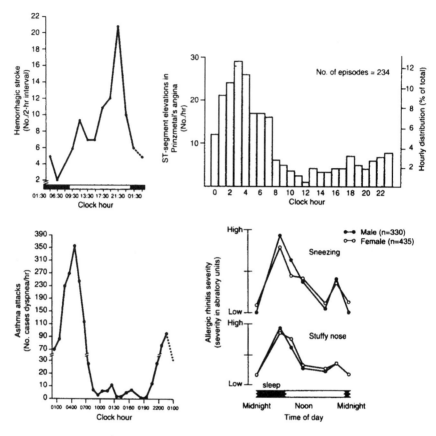

Figure 4: Twenty-four-hour patterns in human disease in diurnally active persons. Upper left: Hemorrhagic stroke is ~threefold greater at 1930 than 0930. Upper right: Episodes of ST-segment elevation in Prinzmetal angina (58 patients; 234 episodes) are ~tenfold more frequent during sleep at 0400 than during activity at 1200. Lower left: Asthma attacks in a group of 3,160 patients are at least 100-fold more frequent at 0400 than midday. Lower right: Allergic rhinitis (sneezing and stuffy nose) in 330 male (closed circles) and 435 female (open circles) patients is worse on morning awakening. (Figure created with permission using graphs from Reinberg and Smolensky,[2] Kuriowa,[34] Dethlefsen and Repges,[27] and Reinberg et al.[39])

- Asthma, with the risk of dyspnea, wheezing, and other symptoms of acute exacerbation being at least 100-fold greater during night-time sleep than during daytime activity;[27]

- Arthritis, with the signs and symptoms of rheumatoid arthritis being two to three times more intense in the morning upon awaking than in the afternoon and those of osteoarthritis being about two times greater around the middle to later portion of the activity period than at other times during the 24 hours;[28,29]

- Peptic ulcer disease, with gastric and duodenal pain at the initial onset of disease or on its recurrence, typically taking place or worsening during the early hours of night-time sleep and common also after gastric emptying following meals;[30]

- Epilepsy, with the occurrence of overt seizures often restricted to particular times of the day or night;[31]

- Migraine, with headache being five to eight times more frequent in the morning when awaking from night-time sleep and during the initial hours of daytime activity than in the afternoon and evening;[32]

- Exertional angina, with ischemic events (chest pain and ST segment depression) being about five times more common during the first hours of the daily activity span than during the afternoon;[33]

- Prinzmetal angina, with episodes of ST-segment elevation being 15 to 20 times more frequent during the sleep than activity span;[34]

- Myocardial infarction, with morbidity and mortality being at least twice as great during the initial hours of the daily activity span than in the evening;[35,36]

- Stroke, with the incidence of the thrombotic form being three times more common at the beginning of the activity period than at night and that of the hemorrhagic form being five to six times more likely late in the evening than overnight and in the morning.[2,37]

Important predictable-in-time alterations in human physiology and disease status also occur over other time domains, such as the week, menstrual cycle, and year.[2]

Biological Rhythms and the Diagnosis of Disease

The conduct and interpretation of diagnostic procedures are often done without awareness of possible chronobiological effects. With reference to the construct of homeostasis, the time when the majority of medical tests are conducted is expected to be of little consequence. This is not always the case. The airway patency of asthma patients varies greatly throughout the 24 hours. It is best during the daytime and during regular clinic hours, and is the poorest at night.[38] The peak expiratory flow (PEF) and 1-second forced expiratory volume (FEV_1), as quantified by spirometry, vary by 25% during the 24 hours even in mildly affected patients and as much as 50% in more severely affected patients. The response of allergic patients to intradermal antigen testing varies threefold on average during the 24 hours; cutaneous reactivity is least in the morning and greatest in the evening.[39] The results of certain laboratory procedures also are affected by the circadian time of their conduct. Cortisol and most other hormones exhibit high-amplitude circadian rhythms; thus, the time when blood samples are withdrawn must be taken into consideration in interpreting laboratory reports.[40] Ambulatory patient-assessment devices, such as Holter and blood pressure monitors (ABPM), clearly reveal prominent 24-hour rhythmicity in endpoints diagnostic of cardiovascular disease.[25,33]

The reference values used in making the differential diagnosis of normotension versus hypertension are derived from daytime blood pressure measurements performed on presumably diurnally active normal persons. Physicians who regularly depend on ABPM assessments for the diagnosis of hypertension know firsthand of the high-amplitude circadian variation in blood pressure and thus have appreciation of the impact of chronobiology on the practice of clinical medicine. Nonetheless, of practical clinical concern is determining how to best utilize and interpret the 24-hour blood pressure data, especially in reference to conventional norms. Twenty-four-hour ABPM provides much more additional information about the patient's blood pressure than does once-a-day office assessment.[22] The clinical implication of new chronobiological endpoints based on around-the-clock ABPM, such as daytime, night-time, and 24-hour blood pressure means; daytime, night-time, and 24-hour blood pressure loads (percent time systolic and diastolic pressures are greater than specified limits); morning blood pressure surge (rate of rise in blood pressure between the termination of nocturnal sleep and commencement of diurnal activity); and presence or absence of blood pressure dipping during sleep, among others, are currently under study. Moreover, knowledge of the circadian patterning of blood pressure raises new questions, not only about the underlying rhythm-dependent mechanisms of hypertension but also about the appropriate pharmacotherapy of its management.[41]

Biological Rhythms and the Behavior of Medications

It is generally recognized that the behavior of various medications can be affected by meal content and timing. It also may be affected by the body's biological time structure. For example, large-amplitude, 24-hour bioperiodicities in gastric hydrogen ion concentration, stomach emptying plus blood flow to the gastrointestinal tract, liver, and kidney can significantly influence the kinetics of medications.[2,42–47] Rhythms in cellular and subcellular functions in drug-targeted tissues can also have a significant impact on the effects of treatments.[2,42,43]

Chronopharmacology is the investigative science concerned with the biological rhythm dependencies of medications.[2,42–47] Chronopharmacological studies have resulted in new insight and concepts germane to clinical medicine.

Chronokinetics refers to rhythm-dependent, administration-time differences in the rate and extent of drug absorption as well as drug distribution and elimination.[2,46,47] Chronokinetic phenomena are often related to the peculiarities of the dosage form and parent drug. The chronokinetics of many different classes of medications have been demonstrated, including benzodiazepines, beta-receptor antagonists, bronchodilators, and nonsteroidal anti-inflammatories (NSAIDs), as well as several other factors, as shown in Figure 5.[45–51]

Chronesthesy refers to rhythm-dependent, administration-time differences in the desired and/or undesired effects of medications.[2,42–44] Chronesthesies result from rhythms in receptor number or conformation, rate-limiting steps in metabolic pathways, and/or the free-to-bond fraction of medications, for example. Chronesthesies are known for analgesics, anticoagulants, beta-receptor antagonists, bronchodilators, corticosteroids, and NSAIDs, among others, as exemplified in Figure 6.[2,49,51–55] Chronesthesies give rise also to rhythms in dose-response relationships as documented for theophylline, beta-receptor agonists, analgesics, and cardiovascular medications.[56]

Chronotoxicity refers specifically to predictable-in-time variation in the undesired effect of medications as a function of their biological time of administration. Very significant circadian chronotoxicities are known, particularly for antitumor agents.[57]

The fundamental strategy underlying the pharmacotherapy of human disease today involves the application of drug delivery technologies and treatment schedules to achieve the homeostatic goal of constancy of blood and tissue drug levels over time. It is assumed that the need for medication by patients is nonvarying throughout the 24 hours, and that constancy of drug level translates directly into constancy in drug effect. Knowledge of day-night and other predictable-in-time variation in the intensity of symptoms or risk of disease coupled with evidence of circadian rhythms in the kinetics, effects, and safety of medications constitute the rationale for a new pharmacological approach to treatment, i.e., chronotherapeutics.

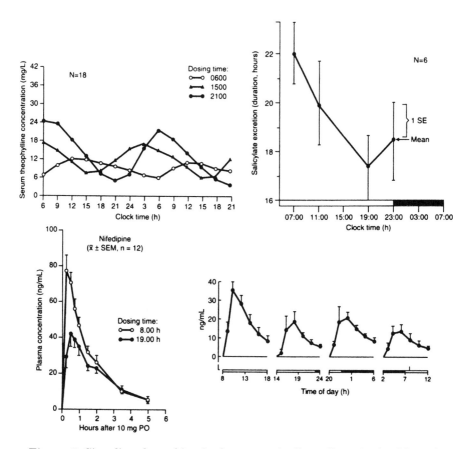

Figure 5: Circadian chronokinetic phenomena in diurnally active healthy subjects. Upper left: Theophylline bioavailability from chronic treatment with 900 mg of Theo-24R (Whitby, USA), an extended-release anhydrous theophylline capsule formulation, in 18 persons is approximately threefold greater with once-daily dosing at 2100 than 0600. Upper right: The clearance of 1 gm aspirin (six persons) is 4.5 hours slower after ingestion at 0700 versus 1900. Lower left: The C_{max} and bioavailability of a single 10 mg oral dose of an immediate-release nifedipine formulation (12 subjects) is significantly greater with dosing at 0800 than 1900. Lower right: The C_{max}, bioavailability and elimination half-life for an 80 mg propranolol dose (four persons) is greatest following an 0800 in comparison to a 1400, 2000, or 0200 administration. Shading along the bottom axes of the individual plots denotes the usual timing and duration of nightly sleep. (Figure created with permission using graphics from Smolensky et al.,[48] Reinberg et al.,[50] Lemmer et al.,[51] and Langer and Lemmer.[49])

Chronotherapeutics:

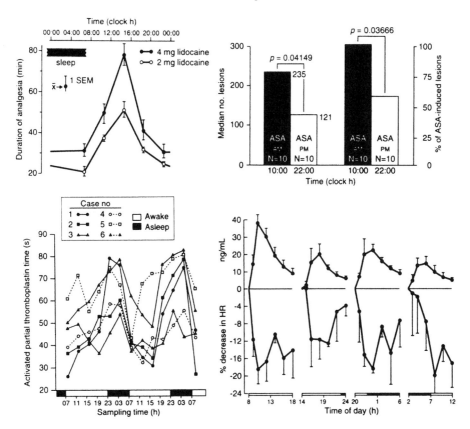

Figure 6: Circadian chronesthesies in diurnally active persons. Upper left: The local analgesic effect of lidocaine (2 and 4 mg dose) in six subjects is at least twofold longer following a 1500 application in comparison to a morning or evening application. Upper right: Hemorrhagic lesions of the gastric mucosa due to 1300 mg aspirin (ASA) ingestion (10 subjects) is twofold greater in number with the 1000 versus 2200 administration. Lower left: Constant rate heparin infusion (six stroke patients) results in cyclic effect with greatest anticoagulant (shortest activated partial prothromblastin time) activity in the afternoon and least such during the night. Lower right: The 80 mg propranolol-induced reduction in sitting heart rate is the same following the 0800 and 0200 oral administrations even though drug C_{max} and bioavailability are unequal, indicating a difference in dose-response. Shading along the bottom axes of the individual plots represents the timing and duration of night-time sleep. (Figure created with permission using graphics from Moore and Goo,[53] Reinberg and Reinberg,[52] Decousus et al.,[54] and Langer and Lemmer.[49])

What, When, and How

Chronotherapeutics is a means of optimizing the effects of medications by delivering different amounts of medication at different times during the 24 hours. Chronotherapeutics takes into account biological rhythms in disease pathophysiology, including temporal patterns of risk and symptom intensity, as well as drug kinetics and dynamics.[2,42–44] It is especially relevant when:

- The risk and/or intensity of diseases and their symptoms vary predictably over time, e.g., 24 hours, menstrual cycle, etc. This is the case, for example, for allergic and infectious rhinitis, angina, arthritis, asthma, migraine, myocardial infarction, stroke, and peptic ulcer disease;

- It is known that the therapeutic-to-toxicity ratio of a medication varies predictably according to chronobiological determinants. This is the case for antitumor medications, synthetic corticosteroids, and NSAIDs, in particular;

- The kinetics and/or dynamics of a medication are known to be biologically rhythm-dependent. This is the case for certain antiarrhythmics, anticoagulants, antihyperlipidemics, antihypertensives, H_1- and H_2-receptor antagonists, and NSAIDs, as well as several others;

- The intent of pharmacotherapy is hormonal substitution to mimic the temporal pattern of healthy individuals. This has entailed synthetic corticosteroids for the treatment of adrenal insufficiency and reproductive hormones for contraception and management of hypothalamic amenorrhea;

- The desired effect of medication can be achieved or optimized only by dosing in a time-modulated manner. This is the case for LHRH administered by ambulatory infusion pumps, for example, to treat hypothalamic amenorrhea.

The morning daily or alternate-day dosing strategy for methylprednisolone, introduced during the 1960s, is considered to be the first chronotherapy used routinely in clinical medicine.[58] Other, often unrecognized, chronotherapies now commonly applied in general medical practice include for day-active patients:

- Evening, once-daily chronotherapy of peptic ulcer disease by H_2-receptor antagonists, and morning chronotherapy of proton-pump antagonists;[59,60]

- Evening, once-daily chronotherapy of asthma with especially formulated theophylline and beta-agonist tablets and capsules;[55,61,62]

- Evening timing of HMGCoA reductase antagonist medication taking into account the circadian rhythm of cholesterol synthesis;[63]

- Unequal, day-night or night-time-only dosing of H_1-receptor antagonist medication to improve the control of the morning symptoms of allergic rhinitis;[64]

- Use of ambulatory, programmable-in-time infusion pumps to clock antitumor medications according to biological rhythms to minimize toxicity and enhance dose-intensities;[57,65]

- Use of programmable-in-time drug delivery devices to deliver tocolytic agents according to the circadian rhythm in uterine contraction to minimize risk of preterm birth;[66]

- Use of programmable-in-time ambulatory infusion pumps to administer LHRH as 90-minute cycles to treat hypothalamic amenorrhea.[67]

- Experimental use of melatonin for resetting the biological clock of travelers and shift workers and the sleep-wake cycle of persons suffering from rhythm-associated sleep disorders.[19]

The Chronobiology and Chronotherapy of Cardiovascular Disease

Day-night patterns in blood pressure, angina, arrhythmias, myocardial infarction, stoke, and other cardiovascular maladies have been quite well documented.[23,25,33–37] In persons adhering to a routine of diurnal activity alternating with nocturnal sleep, the risk of ischemic events is at least two to three times greater during the first hours of daily activity than at night. Day-night patterns in the occurrence and risk of cardiovascular accidents represent the staging of critical endogenous rhythms in relationship to the timing of external triggers during the 24 hours.[23,35,36] The former include circadian variability in sympathetic drive, blood coagulation, blood pressure, coronary blood flow, and myocardial oxygen supply versus demand. Environmental triggers include several morning-time stresses, such as the change from supine to upright posture, the commencement of daily activity, and the sudden onset of mental and emotional loadings.

Presently, management of the circadian rhythm in the risk of ischemic heart disease is based on homeostatic concepts. Compliance to conventional, sustained-released, beta-receptor antagonist, and/or other cardioprotective medications is strongly encouraged, especially at bedtime, with the expectation that therapeutic benefit will persist into the morning. It is now recognized that circadian and other rhythms of the gastrointestinal tract and vital organs are capable of significantly affecting the pharmacokinetics and dynamics of cardiovascular and other medications. This means that the effects of therapeutic interventions administered in identical doses in the morning versus the evening may not be equivalent. Today, once-a-day, ultra-slow-release drug delivery systems are increasingly relied on to attain relatively constant levels of medication throughout the entire 24-hour period. However, even these medications may vary in their behavior with different circadian timings and be less protective than desired during circadian rhythm-dependent times of heightened vulnerability.

Table 2 summarizes the findings of several studies that have addressed administration-time dependencies in the pharmacokinetics of specific antihypertensive and antiarrhythmia medications.[49,68–78] Most investigations compared the kinetics of treatments when administered in the morning at 8 AM versus evening at 8 or 10 PM; only a few explored the kinetics of medications at additional clock times. For the most part, studies relied upon single-dose protocols involving a relatively small number of diurnally active healthy volunteers. Except for the studies with antiarrhythmia agents, statistically significant differences in AUC, C_{max}, and/or $T_{1/2}$ were validated.

Table 3 presents the results of investigations designed to explore possible circadian rhythms in the effects of antihypertensive and antiarrhythmia medications.[49,72,79–90] Trials were conducted on diurnally active healthy volunteers and hypertensive or stable angina patients, using single and chronic dosing protocols. Several studies assessed the effect of equal-interval, equal-dose treatments on the circadian rhythm of blood pressure of essential hypertensive patients. Others compared the effects of morning versus evening treatment on blood pressure control, heart rate, or other variables; only one protocol followed the response of persons to dosing at additional times of the day and night. In

Table 2

Circadian Chronopharmacokinetics of Cardiovascular Medications

Medication, Dose Study Conditions[1,3]	Treatment (Rx) Times	No. and Type Subjects[2]	Major Findings	Reference Author, Yr
SR-Propranolol, 80 mg p.o. (single dose)*	2 & 8 AM, 2 & 8 PM	4H	↑ C_{MAX} & ↓ $T_{1/2}$ with 8 AM Rx	Langner & Lemmer, 1988
Propranolol, 80 mg p.o. (single dose)*	2 & 8 AM, 2 & 8 PM	6H♂	C_{MAX} & AUC lowest for 2 PM vs. other Rx times	Markiewicz et al., 1979
Dipyridamole, 75 mg p.o. (single dose)*	6 AM, 2 PM, 10 PM	6H♂	6 AM Rx AUC 25% > than 10 PM Rx	Markiewicz et al., 1979
IR-Nifedipine, 10 mg p.o. (single dose)*	8 AM, 7 PM	12H♂	AUC & C_{MAX} highest & $T_{1/2}$ lowest for 8 AM Rx	Lemmer, 1991
SR-Verapamil, 360 mg. p.o. (chronic treatment)*	8 AM, 10 PM	10 SAP	AUC highest & T_{MAX} greatest for 10 PM Rx	Jespersen et al., 1989
SR-Enalapril, 10 mg p.o. (3 week treatment)*	8 AM, 8 PM	8 EHP	T_{MAX} for 7 AM Rx 2.1 H < than T_{MAX} 7 PM Rx	Witte et al., 1993
Digoxin, 0.5 mg p.o. (single dose)*	8 AM, 8 PM	9H	Highest $\bar{X}$ drug conc., $T_{1/2}$ & V_D after 8 AM Rx; Highest AUC after 8 PM Rx	Mrozikiewicz et al., 1988
Isosorbide Dinitrate, 20 mg. p.o. (single dose)*	2 & 8 AM, 2 & 8 PM	6H♂	AUC greatest for 2 & 8 AM Rx; $T_{1/2}$ greatest for 8 PM & 2 AM Rx	Blume et al., 1986

Table 2 (*Continued*)

Circadian Chronopharmacokinetics of Cardiovascular Medications

Medication, Dose Study Conditions[1,3]	Treatment (Rx) Times	No. and Type Subjects[2]	Major Findings	Reference Author, Yr
Digoxin, 0.5 mg p.o. (single dose)*	7 AM 7 PM	6 VOL	C_{MAX} after 7 AM Rx twice that after 7 PM Rx	Bruguerolle et al.
SR-Isosorbide-5-mononitrate, 60 mg, p.o. (single dose)	8 AM 8 PM	10♂	No Rx time effect on kinetics; effect only on dynamics	Lemmer et al., 1991
Cibenzoline, 160 mg p.o. (multidose)	8 AM 10 PM	8H	C_{MAX} & T_{MAX} reduced for 10 PM Rx	Brazzell et al., 1985
Procainamide, 500 mg p.o. (multidose)	10 AM 10 PM	8 PVC	No Rx-time dependency on kinetics	Fujimura et al., 1989

[1]**IM** = immediate-release; **SR** = sustained-release dose forms; [2]**H** = healthy subjects; **EHP** = essential hypertensive patients; **SAP** = stable angina patients; **PVC** = premature ventricular arrthymia patients; **VOL** = elderly volunteers; [3]*Statistically significant effect = **P ≤ 0.05.**

Table 3

Administration-Time Dependent Effects of Cardiovascular Medications

Medication, Dose (Study Conditions)[1]	Treatment (Rx) Times	No. and Type Subjects[2]	Major Findings	Reference Author, Yr
Propranolol, 80 mg p.o. (single dose)	2 & 8 AM 2 & 8 PM	4H	↓ HR for 2 & 8 AM Rx> than at 2 & 8 PM Rx; time to E_{MAX} & E_{MAX} at 8 PM & 2 AM Rx > than at 8 AM & 2 PM Rx; Rx-time Δ in dose-response	Langner & Lemmer, 1988
Propanolol, 40 mg p.o. (4 weeks)	QID	9 SAP	Rx more effective on 8 AM & 12 PM ETT than 4 PM one	Joy et al., 1982
Oxprenolol, 344 mg p.o. (6 weeks)	TID	20 EHP	Highly effective in ↓ BP only during daytime. No effect on AM rise in BP; 24 hr BP pattern preserved	Raftery et al., 1981; Raftery, 1983
Atenolol, 100 m p.o. (4 weeks)	OD morning vs. OD evening dosing	5 EHP	Both dosing times ↓ BP only during daytime. No effect on AM rise in BP. No Δs between AM vs. PM Rx; 24 hr BP pattern preserved	Raftery, 1983
Pindolol, p.o. (ng)	NG	12 EHP	Good effect on BP & HR only during daytime. No effect on AM rise in BP; 24 hr BP pattern preserved	Raftery, 1983; Raftery et al., 1981

Table 3 (*Continued*)

Administration-Time Dependent Effects of Cardiovascular Medications

Medication, Dose (Study Conditions)[1]	Treatment (Rx) Times	No. and Type Subjects[2]	Major Findings	Reference Author, Yr
Labetalol, p.o. (varying doses, 6 weeks)	TID	12 EHP	Good 24-hr effect on BP & HR; control of AM rise in BP; 24 hr BP pattern preserved	Raftery et al., 1981; Raftery, 1983
Metoprolol, 100 mg p.o. (9 weeks)	BID	6 EHP	Good 24-hr effect on BP. No control of AM rise in BP; 24 hr BP pattern preserved	Raftery, 1983; Raftery et al., 1981
Nifedipine, 20–60 mg p.o. (6 weeks)	BID	9 EHP	BP effect strongest during daytime. No effect on HR. BP & HR 24 hr pattern preserved. Good effect on AM BP rise	Gould et al., 1982
Verapamil, 120–160 mg p.o. (6 weeks)	TID	20 EHP	BP & HR effect mainly during daytime. Poor effect on AM BP rise; BP & HR 24 hr pattern preserved	Gould et al., 1982
Enalapril, 10 mg p.o. (3 weeks)	7 AM 7 PM	8 HTN	*7 AM Rx:* ↓ daytime BP_{SYS} & BP_{DIA}; no effect on night time BP; *7 PM Rx:* ↓ night time BP_{SYS} & BP_{DIA}; no effect on daytime BP. 7 AM Rx time to E_{MAX} < 7PM Rx	Witte et al., 1993

Drug	Timing	Subjects	Effects	Reference
Combination: 10 mg timolol, 25 mg hyrochlorothiazide, 25 mg amiloride p.o.	OD 8 AM	18 EHP	Good effect on BP day & night; BP & HR reduced most during daytime; BP & HR 24 hr patterns preserved. Moderate effect on AM BP rise	Hornung et al., 1982
Xipamide, 20 mg, p.o. (3 months)	OD	13 EHP	Good 24 hr BP control; effect somewhat stronger during daytime; moderation of AM BP rise; 24 hr BP pattern preserved	Raftery et al., 1981
Potassium, 37 meq iv (1 hr infusions)	12 PM vs. 12 AM	5 H	T-wave elevation > for 0000 Rx & associated with 40% higher ^+K plasma concentration	Moore-Ede et al., 1978
Isosorbital-5-mononitrate, 60 mg p.o. (single dose)	8 AM 8 PM	10 H♂	*8 PM Rx:* time to E_{MAX} on BP_{SYS} (2.8h), BP_{DIA} (2.9h) & HR (3.8h) < *8 AM Rx:* BP_{SYS} (5h), BP_{DIA} (6h) & HR (5.2h). Rx-time Δ in dose-response	Lemmer et al., 1991; Blume et al., 1986
Isosorbide dinitrate, 20 mg p.o. (single dose)	2 & 8 AM 2 & 8 PM	6 H♂	Orthostatic tachycardia ↑ for 4 h after all Rx times except 8 AM; orthostasis > with 2 AM Rx	Lemmer et al., 1991; Lemmer et al., 1986

Table 3 *(Continued)*

Administration-Time Dependent Effects of Cardiovascular Medications

Medication, Dose (Study Conditions)[1]	Treatment (Rx) Times	No. and Type Subjects[2]	Major Findings	Reference Author, Yr
Nitroglycerin, 0.6 mg s.l. (single dose)	5 AM 1–4 PM	13 PVA	*AM Rx*: right/left coronary artery dilitation ↑ 74 ± 40% vs. 12 ± 12% for PM Rx	Yasue et al., 1979
Isradipine, 5 mg p.o. (4 weeks)	8 AM 8 PM	16 HTN-RF	Before Rx, BP nondipping nocturnally; HR ↓ 17.4% from daytime level. *8 AM Rx*: Night BP_{SYS} & BP_{DIA} ↓ 4.8/8.7%; *8 PM Rx*: night BP_{SYS} & BP_{DIA} ↓ 7.5/10.9%	Portaluppi et al., 1995

[1]**P.O.** = oral dosing; **S.L.** = sublingual dosing. [2]**H** = healthy subjects; **HTN** = essential hypertensive patients; **HTN-RF** = hypertensive patients secondary to renal failure; prinzmetal variant angina patients.

studies on patient groups, administration-time-dependencies were detected in heart rate, blood pressure, or coronary artery tone following isosorbital-5-mononitrate, isosorbide dinitrate, nitroglycerin, propranolol, enalapril, and isradipine. Propranolol had greater heart rate depressing effect following morning administration compared to afternoon or evening administration.[49] Morning enalapril dosing was more effective in controlling blood pressure during the day than during the night; evening dosing had a greater effect on night-time than on daytime blood pressure.[72] Also, morning, as opposed to evening, treatment with enalapril resulted in substantially earlier onset of maximum drug effect. Yasue and coworkers[88] demonstrated a circadian rhythm dependency in the effect of nitroglycerin. In Prinzmetal variant angina patients, the medication was several times more effective in decreasing the tone of the large coronary arteries following morning dosing compared to afternoon sublingual dosing. Portaluppi and colleagues[89] demonstrated the differential effectiveness of the dihydropyridine calcium antagonist, isradipine, on sleep-time blood pressure level in nondialysis, chronic renal failure patients. Evening isradipine administration reinstituted the normal nocturnal decline in blood pressure; morning administration was less effective in achieving this. Since elevated night-time blood pressure has been linked to kidney pathology,[91,92] an evening isradipine dosing schedule might enhance the medication's therapeutic benefit.

The findings of several clinical investigations further substantiate circadian rhythm dependencies in the effect of cardiovascular medications. In general, beta-adrenoceptor antagonists exert a greater effect on the heart and vasculature during the day than during the night.[79–81] This is true of both selective and nonselective beta-receptor antagonist medications, whether lipolytic or hydrophilic in form or whether they are eliminated mainly by hepatic metabolism or renal excretion. Examples of circadian rhythm dependencies of adrenoceptor antagonists are shown in Figure 7. The beta-antagonists atenolol, metoprolol, oxprenolol, and pindolol all are highly effective in controlling blood pressure during the day but not at night. The effect of this class of medications appears to be independent of the dosing schedule whether · it is once, twice, or three times daily. Moreover, the effect of atenolol on blood pressure control in essential hypertensive patients is the same whether dosed once daily in the morning at 7 AM or evening at 11 PM.[80,81] Finally, none of the beta-blockers effectively control the morning rapid rise of blood pressure coincident with activity onset. Labetalol, which is thought to exert both alpha- and beta-receptor antagonism, exerts very good 24-hour control of blood pressure, including moderation of the morning rise. The conventional dosage forms of the calcium-channel-antagonists nifedipine and verapamil (Fig. 8), taken

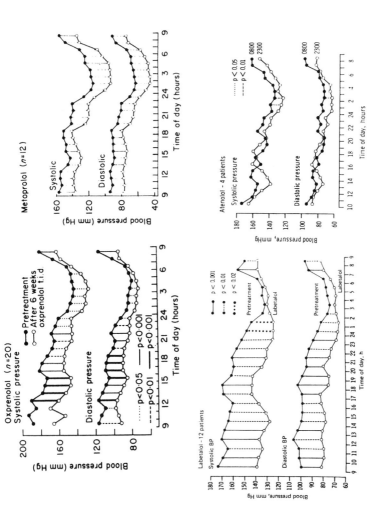

Figure 7: Blood pressure control in essential hypertensive patients by TID oxprenolol (upper left), BID metoprolol (upper right), TID labetalol (lower left) and morning or evening once-daily atenolol (lower right). The first three beta-antagonists preserve the blood pressure pattern; however, their effect is greater during the day than night. These medications, nonetheless, exert rather poor moderation of the morning blood pressure rise. In contrast, labetalol exerts its blood effect both during the day and nighttime with moderation of morning pressures. (Graphs reproduced with permission from Raftery et al., 1981 and 1983.[80,81]

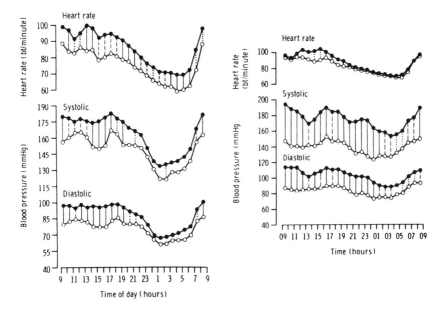

Figure 8: Effect of TID verapamil (left) and BID nifedipine (right) on blood pressure and heart rate of essential hypertensive patients. Verapamil has a stronger effect, particularly on blood pressure, during the day than night-time. Nifedipine has good 24-hour effect on blood pressure, although it is somewhat stronger during the daytime. There is little or no effect of the drug on heart rate. Verapamil dosed at equal intervals has little effect on the morning rise in blood pressure, while nifedipine does. (Graphs reproduced with permission from Gould et al., 1982.[82,83])

two or three times daily in equal doses, also exhibit stronger effect on blood pressure during the day than night.[80–83] Finally, oral nitrates induce greater vasodilatory effect following night-time and early morning compared to afternoon and evening administration.[51,75,87]

To a great extent, the chronopharmacological study of most cardiovascular medications has relied primarily on protocols incorporating only morning and evening dosing times. In several investigations, meals and posture were carefully controlled. However, two time-point study protocols are inadequate to assess circadian rhythm dependencies in the behavior and effect of medications. The particular times selected for dosing may not correspond to those of greatest and least effect on drug kinetics and dynamics. Chronobiologists prefer that studies involve at least three or four different drug administration times to properly explore chronopharmacological phenomena. Nonetheless, circadian rhythms in drug behavior have been substantiated for several classes of cardiovascular medications.

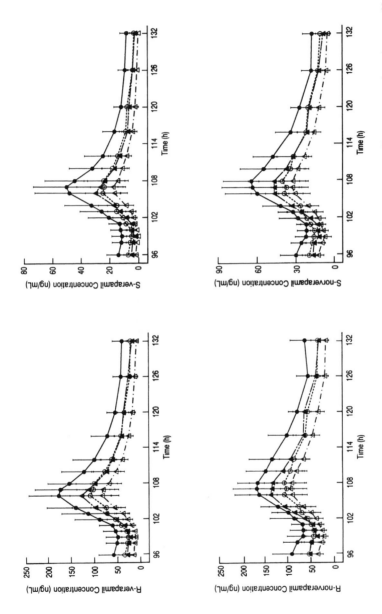

Figure 9: Mean 24-hour plasma R-verapamil, S-verapamil, R-norverapamil and S-norverapamil concentration-time profiles at steady state in young and elderly men and women following 180-mg dosing of Covera-HS[R] (Searle, USA). Open triangle: young men 19–43 years old; open circle: elderly men 65–80 years old; closed triangle: young women 19–43 years old; closed circle: elderly women 65–80 years old. (Reproduced with permission from Gupta et al.[93])

Significant progress has been achieved recently in realizing the chronotherapy of essential hypertension and morning-time angina. In the United States, a novel once-a-day formulation of the calcium antagonist verapamil has been recently devised.[93–96] Physiologically proportioned-release verapamil (Covera-HS; Searle[R], USA) is specifically intended for bedtime dosing. When so dosed, the serum concentration of the drug is held low until the few hours prior to the expected termination of nocturnal sleep. At this time an elevated verapamil blood level is achieved and maintained until the end of the 24-hour dosing interval as shown in Figure 9.[93] The drug delivery program for this medication is specifically designed to moderate the morning surge and daytime elevation of blood pressure of essential hypertensive patients (Fig. 10) as well as reduce the morning risk of ischemic heart disease.[95,96]

Summary

This chapter has identified several clinically significant disparities between the constructs of homeostasis and chronobiology. Many scientists and practitioners are perplexed when first introduced to chronobiology; they have difficulty rectifying the perceived differences between biological constancy and rhythmicity. In reality, they are compatible; the set points at which homeostatic mechanisms are activated undergo precise rhythmic variation throughout the 24 hours, menstrual cycle, and year. Homeostatic mechanisms are responsible for the moment-to-moment regulation of the internal environment, while chronobiological ones are responsible for preparing the organism to cope with predictable-in-time challenges associated with temporal patterns in activity-rest, reproduction, and environment.

Medical chronobiology entails understanding how biological rhythms affect the diagnosis, manifestation, and treatment of human disease. Biological functions are not constant during the 24 hours, menstrual cycle, and year as inferred by the concept of homeostasis. Instead, they vary predictably according to defined, inherited bioperiodicities. The human biological time structure is in part responsible for the observed 24-hour patterns in disease, including cardiovascular ones. The prevention and treatment of cardiovascular disease must take into account chronobiological factors. Circadian rhythms of the gastrointestinal system and vital organs also are capable of affecting the kinetics and dynamics of a great many medications, including those prescribed for the management of hypertension and heart disease. Medical chronobiology seeks to elucidate rhythm determinants of the pathophysiology of disease as well as optimize medications by timings to biological need, which varies rhythmically during the 24 hours. Novel drug delivery

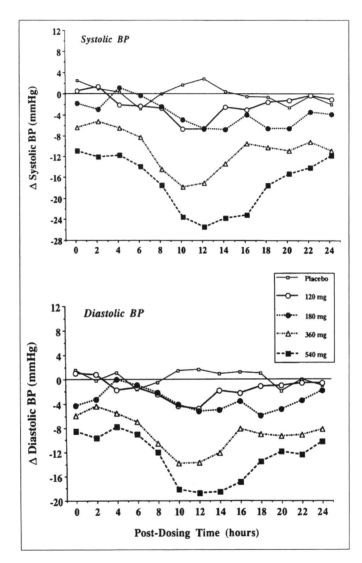

Figure 10: Changes from baseline in 24-hour blood pressure pattern of groups of 55–60 diurnally active essential hypertensive patients treated for eight weeks at 10 PM with placebo or one of four doses of physiological pattern release-verapamil (Covera-HS[R]; Searle, USA). Systolic and diastolic blood pressure shown in the upper and lower portion of the graph. (Reproduced with permission from White et al.[95])

technology now makes possible chronotherapeutic interventions for the management of hypertension and ischemic heart disease.

Chronobiology is of great relevance to the practice of medicine. Clinical medicine generally addresses the questions of: *What* is ailing the patient? *Why and how* should the patient be treated? and *How much* (dose of) medication should be prescribed? Today, answers to other important questions pertaining to *when* must be ascertained: *When* is the risk of disease greatest? *When* are symptoms most troublesome? *When* are diagnostic tests to be conducted? and *When* are treatments to be timed? The past few decades have witnessed very rapid advances in the science of medical chronobiology and chronotherapeutics. These are now being assimilated into clinical medical practice.

References

1. Moore-Ede MC, Sulzman FM, Fuller CA: The Clocks that Time Us. Cambridge, Harvard University Press, 1982.
2. Reinberg A, Smolensky MH: Biological Rhythms and Medicine. New York, Springer Verlag, 1983; p 305.
3. Touitou Y, Haus E (eds): Biological Rhythms in Clinical and Laboratory Medicine. Springer Verlag, Heidelberg, 1992.
4. Halberg F: Chronobiology. Ann Rev Physiol 1975; 31:675–735.
5. Haus E, Touitou Y: Principles of clinical chronobiology. In: Haus E, Touitou Y (eds), Biological Rhythms in Clinical and Laboratory Medicine, Heidelberg, Springer Verlag, 1992; pp 6–34.
6. Halberg F, Carandente F, Cornelissen G, Katinas GS: Glossary of chronobiology. Chronobiologia 1977; 4(Suppl 1):1–189.
7. Thomas V, Chan W, Yetman R, Smolensky MH, Portman RJ: Circadian rhythm of blood pressure in healthy persons and kidney transplant patients: dependence on activity. Chronobiol Intern 1995; 12:419–426.
8. Brown FA Jr: Biological chronometry. Am Natural 1957; 91:129–133.
9. Bunning E: Zur Kenntnis der erblichen Tagesperiodizitat bei den Primarblattern von Phaseolus multiflorus. Jahrb Wiss Bot 1935; 81:411–418.
10. Konopka RJ, Benzer S: Clock mutants of Drosophila melanogaster. Proc Natl Acad Sci USA 1971; 68:2112–2116.
11. Motohashi Y, Reinberg A, Ashkenazi IE, Bicakova-Rocher A: Genetic background of circadian dsychronism: comparison between Asiatic-Japanese and Caucasian-French populations. Chronobiol Intern 1995; 12: 324–332.
12. Aschoff J, Daan S, Groos G (eds): Vertebrate Circadian Systems. Berlin, Springer Verlag, 1982.
13. Ralph MR, Foster RG, Davis FC, Menacker M: Transplanted suprachiasmatic nucleus determines circadian period. Science 1990; 247:975–978.
14. Wever RA: The Circadian System of Man: Results of Experimentation Under Temporal Isolation. Berlin, Springer Verlag, 1979.
15. Halberg F, Simpson H: Circadian acrophase of human 17-hydroxycorticosteroid excretion referred to midsleep rather than midnight. Hum Biol 1967; 39:405–413.

16. Halberg F, Reinberg A, Haus E, Ghata J, Siffre M: Human biological rhythms during and after several months of isolation in underground natural caves. Natl Speleol Soc Bull 1970; 32:89–115.
17. Aschoff J, Wever R: The circadian system of man. In: Aschoff J (ed), Handbook of Behavioural Neurobiology. London, Plenum Press, 1981; pp 311–331.
18. Reinberg A: Chronobiological field studies of oil refinery shift workers. Chronobiologia 1979; 6(Suppl 1):1–119.
19. Arendt J, Aldhous M, Marks V: Alleviation of jet lag by melatonin: preliminary results of controlled double blind trial. Br Med J 1986; 292:1170.
20. Baumgart P: Circadian rhythm of blood pressure: internal and external time triggers. Chronobiol Intern 1991; 8:444–450.
21. Halberg F, Drayer JIM, Cornelissen G, Weber MA: Cardiovascular reference data base for recognizing circadian mesor- and amplitude-hypertension in apparently healthy men. Chronobiologia 1984; 11:275–298.
22. Pickering TG: Ambulatory Monitoring and Blood Pressure Variability. London, Science Press, 1991.
23. Smolensky MH, Tatar SE, Bergman SA, Losman JG, Barnard CN, Dacso CC, Kraft IA: Circadian rhythmic aspects of human cardiovascular function: a review by chronobiologic statistical methods. Chronobiologia 1976; 3:337–371.
24. Portaluppi F, Montanari L, Ferlini M, Vergnani L, D'Ambrosi A, Cavallini AR, Bagni B, degli Uberti E: Consistent changes in the circadian rhythms of blood pressure and atrial natriuretic peptide in congestive heart failure. Chronobiol Intern 1991; 8:432–439.
25. Portaluppi F, Montanari L, Ferlini M, Gilli P: Altered circadian rhythms of blood pressure and heart rate in non-hemodialysis chronic renal failure. Chronobiol Intern 1990; 7:321–327.
26. Reinberg A, Gervais P, Levi F, Smolensky MH, Del Cerro L, Ugolini L: Circadian and circannual rhythms of allergic rhinitis: an epidemiologic study involving chronobiologic methods. J Allergy Clin Immunol 1988; 81:51–62.
27. Dethlefsen U, Repges R: Ein neues therapieprinzip bei nachtlichen asthma. Med Klin 1985; 80:44–47.
28. Kowanko ICR, Knapp MS, Pownall R, Swannel AJ: Domiciliary self-measurement in rheumatoid arthritis and the demonstration of circadian rhythmicity. Ann Rheum Dis 1982; 41:453–455.
29. Levi F, LeLouran C, Reinberg A: Chronotherapy of osteoarthritic patients: optimization of indomethacin sustained release (ISR). Annu Rev Chronopharmacol 1984; 1:345–348.
30. Moore JG, Vener KJ, Szabo S (eds), Chronobiology and Ulcerogenesis. Chronobiol Intern 1987; 4:1–122.
31. Halberg F, Howard RB: 24-hour periodicity and experimental medicine. Postgrad Med 1958; 73:20–32.
32. Solomon GD: Circadian rhythms in migraine. Cleveland Clin J Med 1992; 59:326–329.
33. Rocco MB, Barry J, Campbell S, Nabel E, Cook E, Goldman L, Selwyn AP: Circadian variation of transient myocardial ischemia in patients with coronary artery disease. Circulation 1987; 275:395–400.
34. Kuroiwa A: Symptomatology of variant angina. Jpn Circ J 1978; 42:459–460.

35. Purcell H, Mulcahy D, Fox K: Circadian patterns of myocardial ischemia and the effects of antianginal drugs. Chronobiol Intern 1991; 8:309–320.

36. Ridker PM, Willich SN, Muller JE, Hennekens CH: Aspirin, platelet aggregation and the circadian variations of acute thrombotic events. Chronobiol Intern 1991; 8:327–335.

37. Marshall J: Diurnal variation in the occurrence of strokes. Stroke 1977; 8:230–231.

38. Smolensky MH, Barnes PJ, Reinberg A, McGovern JP: Chronobiology and asthma. I. Day-night differences in bronchial patency and dyspnea and circadian rhythm dependencies. J Asthma 1986; 23:321–343.

39. Reinberg A, Zagulla-Mally Z, Ghata J, Halberg F: Circadian reactivity to rhythms of human skin to house dust, penicillin and histamine. J Allergy Clin Immunol 1969; 44:292–306.

40. Haus E, Touitou Y: Chronobiology in laboratory medicine. In: Touitou Y, Haus E (eds), Biologic Rhythms in Clinical and Laboratory Medicine, Heidelberg, Springer-Verlag, 1992; pp 673–708.

41. Lemmer B: Temporal aspects of the effects of cardiovascular active drugs in humans. In: Lemmer B (ed), Chronopharmacology: Cellular and Biochemical Interactions. New York, Marcel Dekker, Inc. 1989; pp 525–541.

42. Reinberg, AE: Concepts of circadian chronopharmacology. In: Hrushesky WJM, Langer R, Theeuwes F (eds), Temporal Control of Drug Delivery, New York, New York Academy of Science 1991; pp 102–115.

43. Reinberg A, Labrecque G, Smolensky MH: Chronobiologie et Chronotherapeutique. Heure Optimale d'Administration des Medicaments. Paris, Flammarion, 1991.

44. Lemmer B: (ed), Chronopharmacology: Cellular and Biochemical Interactions. New York, Marcel Dekker, Inc., 1989.

45. Reinberg A, Smolensky MH: Circadian changes in drug disposition in man. Clin Pharmacokinet 1982; 7:401–420.

46. Lemmer B, Bruguerolle B: Chronopharmacokinetics. Are they clinically relevant? Clin Pharmacokinetics 1994; 26:419–427.

47. Belanger P: Chronopharmacology in drug research and therapy. Adv Drug Res 1993; 24:1–80.

48. Smolensky MH, Scott PH, Harris RB: Administration-time-dependency of the pharmacokinetic behavior and therapeutic effect of a once-a-day theophylline in asthmatic children. Chronobiol Intern 1987; 4:435–448.

49. Langner B, Lemmer B: Circadian changes in the pharmacokinetics and cardiovascular effects of oral propranolol in healthy subjects. J Clin Pharmacol 1988; 33:619–624.

50. Reinberg A, Clench J, Ghata J, Albuker F, Dupont J, Zagula-Mally ZW: Rythmes circadiens de l'excretion urinaire du salicylate (chronopharmacocinetique) chez l'adulte sain. CR Acad Sci (Paris) 1975; 280:1697–1700.

51. Lemmer B, Scheidel B, Behne S: Chronopharmacokinetics and chronopharmacodynamics of cardiovascular active drugs. propranolol, organic nitrates, nifedipine. In: Hrushesky WJM, Langer R, Theeuwes F (eds), Temporal Control of Drug Delivery. Ann NY Acad Sci 1991; 618:166–181.

52. Reinberg A, Reinberg MA: Circadian changes of the duration of action of local anaesthetic agents. Naunyn Schmiedebergs Arch Pharmacol 1977; 297:149–159.

53. Moore JG, Goo RH: Day and night aspirin-induced gastric mucosal damage and protection by ranitidine in man. Chronobiol Intern 1987; 4:111–116.

54. Decousus HA, Croze M, Levi FA, Jaubert J, Perpoint B, Reinberg A, Queneau P: Circadian changes in anticoagulant effect of heparin infused at a constant rate. Br Med J 1985; 290:341–344.

55. D'Alonzo GE, Smolensky MH, Feldman S, Gianotti LA, Emerson MB, Staudinger H, Steinijans VM: Twenty-four-hour lung function in adult patients with asthma: chronoptimized theophylline therapy once daily in the evening versus conventional twice-daily dosing. Am Rev Respir Dis 1990; 142:84–90.

56. Reinberg A, Levi F: Dose-response relationships in chronopharmacology. Ann Rev Chronopharmacol 1990; 6:25–46.

57. Hrushesky WJM (ed): Circadian Cancer Therapy, Boca Raton, CRC Press, Inc., 1994.

58. Harter JG, Reddy WJ, Thorn GW: Studies on an intermittent corticosteroid dosage regimen. N Engl J Med 1963; 296:591–595.

59. Ireland A, Colin-James DG, Gear P, Golding PL, Romage JK, Williams JG, Leicester RJ, Smith CL, Ross G, Banforth J, DeGara CJ, Gledhill T, Hunt RH: Ranitidine 150 mg twice daily vs. 300 mg nightly in treatment of duodenal ulcers. Lancet 1984; 2:274–276.

60. Chiverton SG, Howden CW, Burget DW, Hunt RH: Omeprazole (20 mg) daily given in the morning or evening: a comparison of effects on gastric acidity, and plasma gastrin and omeprazole concentration. Aliment Pharmacol Ther 1992; 6:103–111.

61. Goldenheim PD, Conrad EA, Schein L: Treatment of asthma by a controlled release theophylline tablet formulation: A review of the North American clinical experience with nocturnal dosing. Chronobiol Intern 1987; 4:397–408.

62. D'Alonzo GE, Smolensky MH, Feldman S, Gianitti L, Emerson M, Gnosspelius Y, Karlson K: Bambuterol in the treatment of asthma: a placebo-controlled comparison of once-daily morning versus evening administration. Chest 1995; 107:406–419.

63. Stalenhoff AFH, Mol MJTM, Stuyt PMJ: Efficacy and tolerability of simvastatin. Am J Med 1989; 87(Suppl 4A): 39s–43s.

64. Reinberg AE: The chronopharmacology of H_1-antihistimines. In: Lemmer B (ed.), Chronopharmacology: Cellular and Biochemical Interactions. New York, Marcel Dekker, Inc., 1989; pp 137–167.

65. Bjarnason G, Kerr IG, Doyle N, Macdonald M, Sone M: Phase I study of 5-fluorouracil and leucovorin by a 14-day circadian infusion in metastatic adenocarcinoma patients. Cancer Chemother Pharmacol 1993; 33:221–228.

66. Lamb F, Gill P, Smith M, Kitzmiller JL, Katz M: Use of the subcutaneous terbutaline pump for long-term tocolysis. Obstet Gynecol 1988; 72:810–813.

67. Gompel A, De Plunkett T, Mauvais-Jarvais P: Induction de l'ovulation par pompe a la LH-RH. Gynecologie 1986; 37:309.

68. Markiewicz A, Semenowicz J, Korczynska J, Boldys H: Temporal variations in the response of ventilatory and circulatory functions to propranolol in healthy men. In: Smolensky MH, Reinberg A, McGovern JP (eds), Recent Advances in the Chronobiology of Allergy and Immunology, Pergamon Press, 1979; pp 185–193.

69. Markiewicz A, Semenowicz K, Korczynska J, Czachowska A: Chronopharmacokinetics of dipyridamole. Int J Clin Pharmacol Biopharm 1979; 17: 222–224.

70. Lemmer B, Nold G, Behne S, Kaiser R: Chronopharmacokinetics and cardiovascular effects of nifedipine. Chronobiol Intern 1991; 8:485–494.

71. Jespersen CM, Frederiksen M, Fisher Hansen J, Klitgaard NA, Sorum C: Circadian variation in the pharmacokinetics of verapamil. Eur J Clin Pharmacol 1989; 37:613–615.

72. Witte K, Weisser K, Neubeck M, Mutschler E, Lehmann K, Hopf R, Lemmer B: Cardio-vascular effects, pharmacokinetics and converting enzyme inhibition of enalapril after morning versus evening administration. Clin Pharmacol Ther 1993; 54:177–186.

73. Mrozikiewicz A, Jablecka A, Lowicki Z, Chmara E: Circadian variation of digoxin pharmacokinetics. Ann Rev Chronopharmacol 1988; 5:457–460.

74. Blume H, Scheidel H, Becker J, Renczes J, Lemmer B: Chronopharmacology of oral isosorbide dinitrate in healthy subjects. Naunyn-Schmiedeberg's Arch Pharmacol 1986; 332:R57.

75. Lemmer B, Scheidel B, Blume H, Becker HJ: Clinical chronopharmacology of oral sustained-released isosorbide-5-mononitrate in healthy subjects. Eur J Clin Pharmacol 1991; 40:71–75.

76. Brazzell RK, Khoo K-C, Schneck DW: Effect of time of dosing on the disposition of oral cibenzoline. Biopharm. Drug Disp 1985; 6:433–440.

77. Fujimura A, Kajiyama H, Kumagai Y, Nakashima H, Sugimoto K, Ebihara A: Chrono-pharmacokinetic studies of pranoprofen and procainamide. J Clin Pharmacol 1989; 29:786–790.

78. Bruguerolle B, Bouvenot G, Bartolin R, Manolis J: Chronopharmacocinetique de la digoxine chez le sujet de plus de soixant-dix ans. Therapie 1988; 43:251–253.

79. Joy M, Pollard CM, Nunan TO: Diurnal variation in exercise response in angina pectoris. Br Heart J 1982; 48:156–160.

80. Raftery EB: The effects of beta blocker therapy on diurnal variation of blood pressure. Eur Heart J 1983; 4(Suppl D):61–64.

81. Raftery EB, Millar-Craig MW, Mann S, Balasubramanian V: Effects of treatment on circadian rhythms of blood pressure. Bio-Telem Pat Monitor 1981; 8:113–120.

82. Gould BA, Hornung RS, Mann S, Balasubramanian V, Raftery EB: Slow channel inhibitors verapamil and nifedipine in the management of hypertension. J Cardiovasc Pharmacol 1982; 4:S369–S373.

83. Gould BA, Mann S, Kieso H, Balasubramanian V, Raftery EB: The 24-hour ambulatory blood pressure profile with verapamil. Circulation 1982; 65:22–27.

84. Hornung RS, Gould BA, Kieso H, Raftery EB: The effect of a combination of timolol hydrochlorothiazide and amiloride on 24 hour blood pressure control using ambulatory intra-arterial monitoring. Br J Clin Pharmacol 1982; 14:415–420.

85. Raftery EB, Melville DI, Gould BA, Mann S, Whittington JR: A study of the antihypertensive action of xipamide using ambulatory intra-arterial monitoring. Br J Clin Pharmacol 1981; 12:381–385.

86. Moore-Ede MC, Meguid MM, Fitzpatrick GF, Boyden CM, Ball MR: Circadian variation in response to potassium infusion. Clin Pharmacol Ther 1978; 23:218–227.

87. Lemmer B, Becker H, Renczes J, Scheidel B, Blume H: On the chron-opharmacology of oral isosorbide dinitrate: pharmacokinetics and cardiovascular effects in healthy subjects. Annu Rev Chronopharmacol 1986; 3:339–342.

88. Yasue H, Omote S, Takizawa A, Nagao M, Miwa K, Tanaka S: Circadian variation of exercise capacity in patients with Prinzmetal's variant angina: role of exercise-induced coronary arterial spasm. Circulation 1979; 59: 938–948.

89. Portaluppi F, Vergnani L, Manfredini R, delgi Uberti EC, Kersini C: Time dependent effect of isradipine on the nocturnal hypertension in chronic renal failure. Am J Hypert 1995; 8:719–726.

90. Lemmer B: Chronopharmacology of cardiovascular medications. In: Reinberg A (ed), Clinical Chronopharmacology: Concepts, Kinetics and Applications. Clinical Pharmacology Monograph Series, No. 4. H-P Kuemmerle (series ed). New York, Hemisphere Pub. Co., 1990; pp 177–204.

91. Timio M, Lolli S, Verdura C, Monarca C, Merante F, Guerrini C: Circadian blood pressure changes in patients with chronic renal insufficiency: a prospective study. Renal Fail 1993; 15: 231–237.

92. Del Rosso G, Amoroso L, Santoferrara A, Fielderling B, Di Liberato, A, Albertazzi A: Impaired blood pressure nocturnal decline and target organ damage in chronic renal failure. J Hypert 1994; 12(Suppl 3): S14.

93. Gupta SK, Atkinson L, Tri T, Longstreth A: Age and gender related changes in steroselective pharmacokinetics and pharmacodynamics of verapamil and norverapamil. Br J Clin Pharmacol, in press.

94. Gupta SK, Yih B, Atkinson L, Longstreth J: The effect of food, time of dosing, and body position on the pharmacokinetics and pharmacodynamics of verapamil and norverapamil. J Clin Pharmacol, in press.

95. White WB, Anders RJ, MacIntyre, Black HR, Sica DA, the Verapamil Study Group: Nocturnal dosing of a novel delivery system of verapamil for systemic hypertension. Am J Cardiol 1995; 76:375–380.

96. Cutler NR, Anders RJ, Jhee SS, Sramek JJ, Awan NA, Bultas J, Lahiri A, Woroszylska M: Placebo controlled evaluation of three doses of a controlled-onset, extended-release formulation of verapamil in the treatment of stable angina pectoris. Am J Cardiol 1995; 75:1102–1106.

11

Therapeutic Considerations for Circadian Pattern of Cardiovascular Events

Charles R. Lambert, MD, PhD,
Prakash C, Deedwania, MD

Introduction

Previous chapters in this volume detail the important and expanding collection of cardiovascular events that follow a circadian or similar pattern of occurrence. These include hypertension, myocardial ischemia, acute myocardial infarction, sudden cardiac death, a variety of cardiac dysrhythmias, and cerebrovascular accidents, among others. The basic mechanisms felt to be responsible for these clinical events or syndromes have been covered in their respective chapters and many of these mechanisms overlap or act in concert in different clinical scenarios. Indeed, it may appear difficult to design therapeutic approaches to prevent or modify cardiovascular events that are so multifactorial in mechanism and that possess heterogeneous time-varying properties as well. This process, or the application of chronotherapeutics to cardiovascular medicine, has also been introduced (see Chapter 10). Certain principles of practical use to the clinician can be distilled from the large and heterogeneous pool of information dealing with time-varying cardiovascular events detailed with this volume.

From: Deedwania PC (ed): *Circadian Rhythms of Cardiovascular Disorders.*
©Futura Publishing Co., Inc., Armonk, NY, 1997.

When a cardiovascular event or syndrome is shown to exhibit circadian or other time-varying characteristics, an essential element for further investigation is a reliable indicator or variable for quantification. If an event such as myocardial infarction is of interest, simple tabulation of incidence may suffice. Mechanisms may be explored by looking at other time-varying indicators such as fibrinolytic activity; however, quantification of the event itself is relatively straightforward. Other tools, such as ST segment monitoring for ambulatory ischemia, may not be so straightforward. Technical aspects of recording, uncertainties about sampling error dependent on the length of time over which measurements are made, and critical interrelated variables such as heart rate and blood pressure make interpretation of information difficult. Therapeutic responses may also be difficult to interpret if the precise response of multiple interrelated variables is either inadequately measured or unknown.

Provided that the event of interest can be quantified appropriately, the chronobiology of that event for normal individuals should be described before attempting to describe the pathophysiological state. Rigorous mathematical characterization may be applied. Proposed mechanisms that may be responsible for the pathophysiology of the event might then be specifically targeted by a therapeutic agent. An effect or effects on the variable or event of interest is quantified. An assessment also needs to be made for time-varying or static adverse effects that may occur due to the therapeutic agent. This process can then be repeated or modified to test different or interrelated mechanisms and explore other time-varying events.

In theory, such a process of investigation and description might lead to better therapeutic modalities for treatment of time-varying events. Although it would appear that the most valuable information might come from such a rigorous, mechanism-driven approach, useful information may also be gained from simple observation of time-varying effects of an effective agent on clinical outcome or a variable of interest. Both the rigorous, mechanism-based chronotherapeutic approach to drug design and therapy as well as simple observation of cause and effect have yielded valuable information. It also appears that the area of precision chronotherapeutics holds great promise for prevention of significant morbidity and mortality that has escaped conventional therapeutic approaches. Such issues are important to understand if accurate assessments of chronotherapeutic efficacy are to be made. Indeed, these considerations are important for the clinician to understand because circadian variation in cardiovascular disease has evolved from scientific curiosity to an important principle for optimal patient care. With these considerations in mind, in this final chapter we wish to examine some practical therapeutic considerations related to circadian patterns of cardiovascular disease and currently available therapeutic agents.

Beta-Adrenergic Blockers

As previously detailed, beta blockers modify many of the possible pathophysiological mechanisms for adverse cardiovascular events probably through decreasing sympathetic tone and subsequent alteration of heart rate and blood pressure. Indeed, beta blockade titrated to optimum therapeutic effect in patients with coronary artery disease abolishes the circadian variation of heart rate.[1] Lesser degrees of beta blockade may ameliorate ischemia without totally abolishing circadian variation. The exact degree to which circadian variation of heart rate or blood pressure should be suppressed in patients with hypertension or ischemic heart disease is unknown. It appears logical that therapy should be targeted to lessen the effect of physiological events that occur in the morning hours; however, the optimal chronotherapeutic regimen for this effect has yet to be defined. After therapy with beta blockade intense enough to abolish circadian variation in heart rate, we have found remaining ischemia to have different temporal characteristics than those seen at baseline. The bandwidth or spectral characteristics of this remaining ischemia after beta blockade suggests a different mechanism such as alteration in coronary blood flow. If this is in fact the case, then proper chronotherapy might require simultaneous therapy with agents of different classes such as a beta blocker and calcium blocker or nitrate. This idea is also supported by analysis of subsets of ambulatory myocardial ischemia based on heart rate activity as done by the Angina and Silent Ischemia (ASIS) Study Group.[2] Evaluation of baseline heart rates and patterns of change in heart rate revealed that, although most episodes of ischemia are preceded by a significant period of increased heart rate, a minority of episodes are not. The latter appeared to be more effectively reduced by nifedipine.

Beta-blocker therapy may also modify the chronobiology of acute myocardial infarction.[3] Within the 1,741 patients included in the Intravenous Streptokinase in Acute Myocardial Infarction Study (ISAM), the only subgroup without a morning increase in the incidence of infarction consisted of patients taking beta blockers before the event. Autonomic disturbances in diabetics also appear to blunt the morning peak of acute myocardial infarction.[4] In a review of 10,791 patients treated in a single center, beta blockade blunted the circadian pattern of acute myocardial infarction but only if the drug was of the selective variety.[5] These observations suggest that timing of administration of beta blocker or other cardioprotective therapy may be critical for prevention of myocardial infarction.

Beta blockade with propranolol in the Cardiac Arrhythmia Suppression Trial (CAST)[6] blunted the morning increase in sudden cardiac death and the peak in ventricular dysrhythmias. Aronow and coworkers also found that the circadian variation of sudden cardiac death and fatal myocardial infarction was abolished by therapy with propranolol in elderly patients with complex ventricular dysrhythmias.[7] The mechanisms by which beta blockers exert these effects obviously may be complex and multifactorial as covered in detail elsewhere. It is important to note that chronotherapeutic considerations for beta blockade, particularly in the setting of ventricular dysrhythmias, may be particularly complex. The importance of this is highlighted by the finding of dissimilar effects of nadolol on power spectra of heart rate and QT intervals.[8] Such observations suggest that therapeutic considerations for electrophysiological events with circadian rhythmicity might require accounting for differential autonomic modulation of heart rate and ventricular repolarization for optimal efficacy.

In all of the various cardiovascular states with a circadian variation in intensity or incidence for which beta blockade is effective, the predominant mechanism, as mentioned earlier and covered elsewhere in this volume, is related to sympathetic nervous system activity. Other mechanisms such as platelet aggregability,[9] although important, may not play as significant a role in myocardial ischemia.[9] From a practical standpoint, it is important for the clinician to realize that a major mediator of sympathetic activity and cardiac autonomic control is mental stress throughout the day.[10] Although the hemodynamic effects of mental stress may be modified by beta blockade and sophisticated chronotherapeutic approaches may optimize control, in many cases therapeutic strategy might also be directed toward alleviation of mental stress itself.

Calcium Antagonists

Chronotherapeutic considerations and results thus far with calcium antagonists have been much more variable than for beta blockers. This is probably due to the obvious greater heterogeneity in basic pharmacology of drugs of this class as well as dosing and mechanistic variables. There appears also to be more variability in correlation between circadian effect and clinical effect with calcium antagonists in certain circumstances when compared to results with beta blockers. For example, ambulatory ischemia over a 5-day monitoring period using diltiazem in a clinically effective dose had little effect on circadian characteristics of heart rate and ST depression other than some lowering of mean and peak levels[11] (Fig. 1). Another study in which amlodipine and diltiazem were

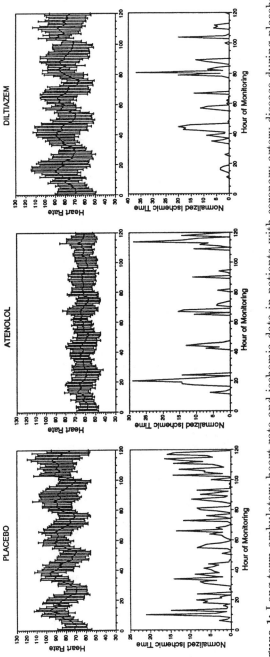

Figure 1: Long-term ambulatory heart rate and ischemia data in patients with coronary artery disease during placebo, atenolol (200 mg qd), and diltiazem (120 mg tid). (Adapted with permission.[11])

compared revealed no blunting of circadian increases in ischemia with either drug, despite lowering of heart rate with diltiazem.[12]

Nifedipine was shown to exhibit no effect on circadian variation of myocardial ischemia at lower doses (30–60 mg/day); however, it was effective at 80 mg/day.[13,14] Use of long-acting or extended-release nifedipine has been shown to blunt the circadian pattern of myocardial ischemia, probably because of improved drug levels and smooth action of this formulation.[15] The effect was similar whether the nifedipine was administered in the morning or evening; however, the addition of beta blockade achieved a much greater therapeutic response.

The beneficial effect of nifedipine on the ischemia that persists after beta blockade suggests that chronotherapeutic considerations for ischemia with calcium antagonists should be different than those for beta blockers.[2] It appears that beta blockade should indeed be targeted to account for circadian characteristics that are predominately sympathetic in etiology. The ischemia remaining after beta blockade appears sensitive to calcium antagonists; however, pharmacodynamic requirements for such ischemia are different and must be optimized for ischemia with a different bandwidth and due to a different mechanism.

Although prospective data are lacking, the circadian distribution of acute myocardial infarction in a large retrospective analysis was not affected by therapy with calcium antagonists.[5]

Thrombolytic and Antiplatelet Therapy

Several studies have shown that the efficacy of thrombolytic therapy in acute myocardial infarction exhibits a circadian pattern (see Chapter 6). Kurnik (Fig. 2) reviewed 692 patients for angiographic patency of the infarct-related artery and found greater patency when t-PA was administered between noon and midnight.[16] A smaller study revealed a similar circadian variation for t-PA that was not seen for streptokinase.[17] The mechanisms that may be important with regard to this difference in therapeutic efficacy in acute myocardial infarction are covered in Chapter 6. The observation itself not only has implication for chronotherapeutics, but also for cost benefit issues and optimization of alternative therapies for acute myocardial infarction such as primary angioplasty.

The effect of aspirin on the circadian pattern of acute myocardial infarction was defined in the Physicians Health Study.[18] Using a placebo-controlled, double-blind design, one aspirin (325 mg) every other day was shown to significantly blunt the circadian variation of myocardial infarction onset.

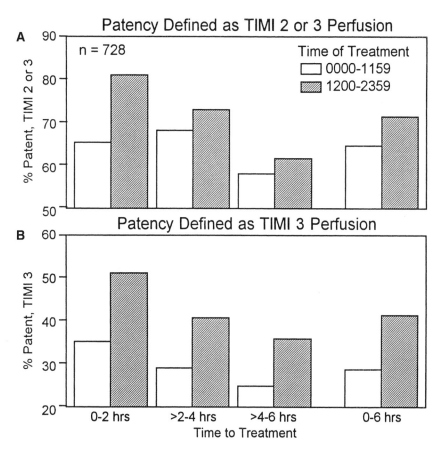

Figure 2: Patency 90 minutes after initiation of thrombolytic therapy as a function of time of treatment for the entire group and as a function of different intervals of time to treatment. **(A)** Patency is defined as TIMI 2 or 3 perfusion. There is greater patency for a shorter time to treatment, $P<0.009$. There is a trend toward greater patency for treatment after noon (1200–2359) for the entire group ($P = 0.07$) and for the subset treated within 2 hours of symptoms ($P = 0.055$). **(B)** Patency is defined as TIMI 3 perfusion. TIMI 3 patency is substantially greater for treatment after noon (1200–2359) $P<0.001$. The number inside each bar indicates the number of patients initiating treatment during that time block. (Used with permission.[12])

Conclusions

The application of therapeutics to chronobiology ranges from simple principles, such as administration of an aspirin every other day to prevent myocardial infarction in the morning, to complex ones, such as designing the optimal combination of agents to tailor precise harmony between ventricular refractoriness, heart rate, blood pressure, and coronary dilation in a patient with ischemic heart disease, malignant dysrhythmias, and hypertension. Many basic questions need to be answered before the field of chronotherapeutics in cardiovascular disease can be expected to produce significant results. For example, although we know hypertension is bad and treatment is good, the proper circadian manipulations needed in various hypertensive patient groups are unknown and will require a great deal of investigation. Is our target to ablate all circadian variation in heart rate and blood pressure for patients with ischemic heart disease? What should the 24-hour profiles of blood pressure and heart rate look like in a 24-year-old with essential hypertension only versus a 58-year-old with hypertension and coronary artery disease with normal ventricular function versus a 40-year-old with hypertension and a dilated cardiomyopathy? Answers to the very precise questions posed by the field of chronotherapeutics in cardiovascular disease will require many highly controlled, prospective investigations based on defined pathophysiology and should not only be focused on efficacy, as most studies to date have been, but should also include evaluation of hard endpoints such as mortality and related outcome indices. Despite these lofty goals, however, much useful and practical information has been gained from observation of circadian variation of cardiovascular events and the effects of various treatment strategies. The physician treating patients with cardiovascular disease can use this information to optimize therapy now, even as newer approaches evolve in the future.

References

1. Lambert CR, Coy K, Imperi G, et al: Influence of beta adrenergic blockade defined by time series analysis on circadian variation of heart rate and ambulatory myocardial ischemia. Am J Cardiol 1989; 64:835–839.
2. Andrews TC, Fenton T, Toyosaki N, et al for the Angina and Silent Ischemia Study Group (ASIS): subsets of ambulatory myocardial ischemia based on heart rate activity. Circadian distribution and response to anti-ischemic medication. Circulation 1993; 88:92–100.
3. Willich SN, Linderer T, Wegscheider K, et al, and the ISAM Study Group: Increased morning incidence of myocardial infarction in the ISAM study: absence with prior β-adrenergic blockade. Circulation 1989; 80:853–858.
4. Tanaka T, Fujita M, Fudo T, et al: Modification of the circadian variation of symptom onset of acute myocardial infarction in diabetes mellitus. Cor Artery Dis 1995; 6:241–244.
5. Hansen O, Johansson B, Gullberg B: Circadian distribution of onset of acute myocardial infarction in subgroups from analysis of 10,791 patients treated in a single center. Am J Cardiol 1992; 69:1003–1008.
6. Peters RW, Mitchell LB, Brooks MM for the CAST investigators: Circadian pattern of arrhythmic death in patients receiving encainide, flecainide or

moricizine in the cardiac arrhythmia suppression trial (CAST). J Am Coll Cardiol 1994; 23:283–289.

7. Aronow WS, Ahn C, Mercando AD, et al: Circadian variation of sudden death or fatal myocardial infarction is abolished by propranolol in patients with heart disease and complex ventricular arrhythmias. Am J Cardiol 1994; 74:819–821.

8. Sarma JS, Singh N, Schoenbaum MP, et al: Circadian and power spectral changes of RR and QT intervals during treatment of patients with angina pectoris with nadolol providing evidence for differential autonomic modulation of heart rate and ventricular repolarization. Am J Cardiol 1994; 74:131–136.

9. Willich SN, Pohjola-Sintonen S, Bhatia SJS, et al: Suppression of silent ischemia by metoprolol without alteration of morning increase of platelet aggregability in patients with stable coronary artery disease. Circulation 1989; 79:557–565.

10. Sloan RP, Shapiro PA, Bagiella E, et al: Effect of mental stress throughout the day on cardiac autonomic control. Biol Psychol 1994; 37:89–99.

11. Lambert CR, Raymenants E, Pepine CJ: Time-series analysis of long-term ambulatory myocardial ischemia: effects of beta-adrenergic and calcium channel blockade. Am Heart J 1995; 129:677–684.

12. Mulcahy D, Purcell H, Sparrow J, et al: Effects of amlodipine versus diltiazem on morning peak in myocardial ischemic activity in angina pectoris. Am J Cardiol 1993; 72:1203–1206.

13. Mulcahy D, Cunningham D, Crean P, et al: Circadian variation of total ischaemic burden and its alteration with anti-anginal agents. Lancet 1988; i:755–759.

14. Nesto RW, Phillips RT, Kett KG, et al: Effect of nifedipine on total ischemic activity and circadian distribution of myocardial ischemic episodes in angina pectoris. Am J Cardiol 1991; 67:128–132.

15. Parmley W, Nesto R, Singh B, and the N-CAP Study Group: Attenuation of the circadian patterns of myocardial ischemia with nifedipine GITS in patients with chronic stable angina. J Am Coll Cardiol 1992; 19(7):1380–1389.

16. Kurnick PB. Circadian variation in the efficacy of tissue-type plasminogen activator. Circulation 1995; 91(5):1341–1346.

17. Becker RC, Corrao JM, Baker SP, et al: Circadian variation in thrombolytic response to recombinant tissue-type plasminogen activator in acute myocardial infarction. J Appl Cardiol 1988; 3:213–221.

18. Ridker PM, Manson JE, Buring JE, et al: Circadian variation of acute myocardial infarction and the effect of low-dose aspirin in a randomized trial of physicians. Circulation 1990; 82(3):897–902.

Index

ACE. See Angiotensin-converting enzyme
Acute myocardial infarction, 1–23
Age and aging, 12–14, 58
Allergic rhinitis, 179, 187
Angina, 181
Angiotensin, 54–55
Angiotensin-converting enzyme (ACE), 54–55, 61
Antiarrhythmic drugs, 155–159
Antiplatelet therapy, 212–213
Arrhythmias
  patterns of, 129–135
  sleep and, 133–134
  sudden cardiac death and, 142
  supraventricular, 71, 135–138
  ventricular, 71, 80–81, 130, 138–142
Arthritis, 180–181
Aspirin, 7–8, 37–38, 212
Asthma, 180, 187
Atenolol, 191, 196
Atherosclerosis, 57
Atrioventricular (AV) node, 71–72, 73, 77

Autonomic nervous system, 132
AV node. See Atrioventricular (AV) node

Beta Blocker Heart Attack Trial (BHAT), 116, 139, 155–156, 160
Beta blockers, 8–9, 37, 155–156, 195, 209–210
BHAT. See Beta Blocker Heart Attack Trial
Biological rhythm. See Chronobiology
Blood pressure, 100, 101, 104, 175, 177, 178, 179, 195, 200

CAD. See Coronary artery disease
Calcium antagonists, 210–212
Calcium channel blockers, 10–11
Cardiac Arrhythmia Suppression Trial (CAST), 115, 155–159, 210
Cardiac repolarization, 78–90
CAST. See Cardiac Arrhythmia Suppression Trial

217